Climate and Code

Monica C. Sanders

Climate and Code

Building Digital Justice in a Connected World

palgrave
macmillan

Monica C. Sanders
Georgetown University
Washington, DC, USA

ISBN 978-3-032-28031-2 ISBN 978-3-032-28032-9 (eBook)
https://doi.org/10.1007/978-3-032-28032-9

Cover credit: weerapatkiatdumrong

This Palgrave Macmillan imprint is published by the registered company Springer Nature Switzerland AG
The registered company address is: Gewerbestrasse 11, 6330 Cham, Switzerland

This book is dedicated to the communities that shared their space, time, and wisdom with me. It is also dedicated to the talented people who work alongside me. Your patience, care, and encouragement made this work possible for all of us.

Foreword

This important book is a vibrant testament to how Monica Sanders creates new horizons. Her work gathers individuals and entities from separate and even disparate sectors, who operate with different priorities, vocabularies, and functions, to newly see and understand lived experiences at the margins of the intersection of climate change and digital divides. The realizations and resolve that form in these new coalitions are energized by innovative ways of achieving solutions and girded by legal and policy-making expertise to create sustained positive impact on a scale that extends from the local to the global. New horizons reorient their viewers to perceive landscapes and aspirations differently; in the work that Monica Sanders and the many partners she inspires achieve. That reorientation has meant calls to action that have been answered by policy changes and concrete transformations in the lives of the people and the world for which she advocates.

In this book, you will find both in-depth analyses of the historical, legal, industrial, political, environmental, technological, and sociological conditions for the climate duress and digital divide experienced by marginalized communities, as well as applicable recommendations for coalition-building, policy-making, understanding across sectors, thinking within intersecting systems, change management, and the reconfiguration of what have been deemed intractable problems. The manifestations of these analyses and recommendations are visible in the creation of communities and the co-creation of solutions forged from climate disasters and digital inequities. I have witnessed firsthand the ability to create community that Monica Sanders practices authentically and meaningfully. Her presence at Grinnell College as the inaugural Social Innovator-in-Residence (another horizon expanded) brought together farmers from surrounding Poweshiek County, impacted agriculturally and financially by climate change and lack of access to the digital grid, with college students based in Grinnell, Iowa, who themselves brought climate and digital experiences from all fifty states and over sixty countries around the world. The resulting mapping and systems work created a community on entirely new terms and with a purpose that continues to be renewed in the work perpetuated and inspired by the internationally recognized Undivide Project and its generative coalitions.

True to its meaning, the Undivide Project and the case studies discussed in this book use language to form the bridge between sectors divided by different goals and vocabulary. The digital sector, now supercharged by artificial intelligence, considers networks, infrastructure, profits, information, and a myriad of other issues shaped by what we now call "technology." The environmental sector, now managing ever-greater impacts of atmospheric and planetary change, operates within global systems and local effects, disasters and recovery, insurance and the economy, energy and resources, and multiple matters that impact human dignity, shaped by what we now call "climate." The two sectors become intertwined within the language of justice and considerations of inequity in both; they speak to each other across the language of resilience and the use of networks in both; they share priorities in the governance of digital and environmental policy; they can further their aspirations in the adaptation upon which both technology and climate insist. And so, while environmental justice might be a familiar term, this book allows a full exploration of its intersection with digital justice, not a new term in our public sphere but one newly valorized and articulated in this book. We need the language offered in this study to advocate for vital new concepts such as "digital equity" in the realm of technology, and "inclusive resilience" in that of climate change.

Working with practitioners is a key aspect of the authenticity of Sanders' work and mobilization. People on the ever-shifting ground of climate disasters have much to share about climate resilience and what makes it possible with those far from those sites in the federal government, where much policy-making occurs, or in the technology industry, where digital networks are laid to make the Internet accessible. Policies and Internet access impact the ability of a community to rebuild in complex and far-reaching ways that are delineated and made understandable in this book, an important disruption to political gridlock and community disempowerment. The immediacy of practitioners who are responding to crises in the moment must be layered on a historical understanding of policies that have created the inequities, in housing, in healthcare, and in Internet access, that this book provides strategies to redress. That we must work both in the immediacy of crisis and response and in the analysis of history and the law, on a scale from the local to the global, is both the ambition and the calling of this book.

The environment, especially in its manifestation in climate change, and technology, especially in its manifestation in generative artificial intelligence, are each often presented as both urgent and overwhelming. The courage of this book is to engage both of these domains and provide recourse to create change by focusing on laws and policies as concrete and accountable practices in society, economy, and government. Further, linking the law to justice underscores the purpose of policy and creates a framework for the work of both. The gift of this book is that the reader is in the good company of the communities and changemakers that Monica Sanders has fostered and with whom she has created new conditions of possibility and "community sovereignty," using law and policy to tackle climate and digital inequity and build climate resilience and digital justice.

Climate and Code ends with a guide to action by calling upon the democratic practice of the social contract. While terms like Internet technology and climate change were not part of Jean-Jacques Rousseau's vocabulary when he wrote The Social Contract in 1762, he and his readers cared deeply about information and its diffusion, and had many emerging opinions on land ownership, human dignity, and the natural world. Calling for a "new social contract," Sanders takes the many insights of this book into the realm of political philosophy and renders the transformative recommendations therein into not only possible manifestations of the "general will" of the social contract but also ready demonstrations of collective civic action. Readers hold in their hands a manifesto of what is possible at the intersection of climate change and the digital divide when a courageous author valorizes the experiences of communities and practitioners to apprehend what has been deemed overwhelming and ungovernable with the resolve and accountability of law and policy. To come to know the work of Monica Sanders is a revelation and a privilege; to carry it forward is a purpose and an honor.

Anne F. Harris
President
Grinnell College
Grinnell, IA, USA

Preface

At the heart of this book lies a vision born from my experiences with The Undivide Project, an initiative I founded in 2022 to champion the cause of universal Internet access. This vision centers on the belief that the Internet's myriad benefits should be accessible to all in an environmentally sustainable and equitably distributed manner. More importantly, this book posits that the intertwined challenges of climate change, digital disparity, and economic instability necessitate a unified response, leading us toward a more just and inclusive world.

The narrative of this book is anchored in the historical context of divestment and disenfranchisement. In the U.S., the legacies of slavery, redlining, and Indigenous displacement, among others, have institutionalized inequalities that persist across generations. This pattern of divestment, rooted in prejudicial policies, has resulted in skewed investment in infrastructure and opportunities, leaving marginalized communities behind. Globally, the scars of colonialism and other historical injustices mirror these challenges, creating a fabric of global inequity that exacerbates vulnerabilities to climate change and digital exclusion.

Through the lens of redlining, this book explores how historical injustices manifest in contemporary challenges such as urban heat islands, environmental justice concerns, and digital deserts. Similarly, the plight of the Global South, grappling with climate adversities, underscores the global nature of these challenges. The narratives of injustice, thus, form a compelling case for reimagining our approach to digital and environmental equity. These historical injustices are not abstract; they persist in the lived experiences of communities like those in West Baltimore and my hometown of New Orleans. The enduring effects of redlining, digital deserts, and environmental degradation are evident in these places, where disparities in access to resources are starkly visible.

When we talk about global inequality, climate change, the digital divide, and how to achieve a more just and inclusive world, one of the key concepts that emerges is digital justice. As I've researched, worked with communities, and engaged in policy work, digital justice has become central to my focus, hence its inclusion in the title of this book. It's a phrase I use uniquely and have expanded on in my research, talks, and work. Our society is becoming increasingly digitized. Services, economic opportunities, education, disaster mitigation, and communication about risks are all shifting to digital spaces. For those who have historically faced barriers to economic opportunities, education, and access to public services like health, housing, and food, these challenges persist in those digital spaces. The reason for this is that we place digital public benefits in the same locations as other public benefits. If you're from a rural, southern, and impoverished community or an urban area that was redlined, you're likely to have limited access to those benefits. In an increasingly digital society driven by artificial intelligence and beyond, this lack of access puts people at an even greater disadvantage.

Some of my colleagues think I coined the term "digital justice." While I may have developed a unique way of using it, I've seen others refer to it as well, often connecting it to civil rights or framing the digital divide in terms of access to hardware and affordability. When I speak of digital justice, I'm referring to a more holistic approach. I focus on the connections between the digital economy and who gets access to the highest levels of financial, social, and educational opportunities within it. I also address how this economy relies on infrastructure that demands large quantities of water, rare earth minerals, and other environmentally impactful resources that are often located in historically divested communities.

For example, cities like Newark and Baltimore, some of the most heavily redlined areas in the country, still carry the scars of their industrial past. Predominantly Black, Latino, and Native American communities in these areas are home to chemical plants, lead manufacturing sites, and cement factories. During the pandemic, people in these communities were labeled "essential workers," often working gig or hourly jobs that paid some of the lowest wages in the country. They didn't have the luxury of working remotely because they lacked access to the necessary hardware, Internet infrastructure, and digital skills. Meanwhile, remote work offered protection and higher income opportunities to those with better access to these resources. This is a clear example of digital injustice, where the same communities that bear the environmental and economic burdens are denied access to the opportunities that digital economies can provide.

Globally, this injustice is mirrored in places like Indonesia, a country with significant political and economic inequalities. It is hyper-exposed to climate risks, including sea-level rise, tsunamis, and volcanic activity, while also being home to extractive industries that further degrade the environment. Many locals are forced into subsistence living, while expats flock to Bali to enjoy the benefits of a digital nomad lifestyle. This stark contrast reflects the broader challenges of digital justice and climate justice on a global scale.

Digital justice, as I define it, goes beyond access to the Internet or technology; it encompasses the intersection of digital economies, environmental resources, and historical inequities. The idea is not just about civil rights but about addressing the deeper, structural inequalities that prevent marginalized communities from fully participating in the digital future. One of the inspirations for this book stems from my work in Baltimore, which is covered in more detail in Chapter Six, which focuses on community solutions.

Before we arrived at solutions and one of the partnerships I'm most proud of, it began with examining a community grappling with the aforementioned legacies of redlining. West Baltimore has a significant number of abandoned homes, some of the highest levels of digital inequity in the city, and high crime rates. The children in this area report poverty, hunger, and trauma at rates higher than in many other places in the country. When I first visited, what struck me was the strength and pride of the community. The resilience I observed was not built with adequate infrastructure and resources. It was forged from repeated exposure to hardship, where people learn to fend for themselves without the necessary support. Despite these challenges, many elders, returning citizens, and lifelong residents maintained a deep sense of connection and pride in their community, embodying what academic researchers call "social cohesion." This neighborhood also bore visible signs of climate and disaster risks, such as watermarks from pollution-contaminated floodwaters on homes. It is also evident that urban heat island effects plagued the area due to insufficient green spaces and tree canopies.

While working on an Internet-related project with colleagues, we took a walk around this community. As we made our way, we saw remnants of failed Wi-Fi initiatives, or good ideas that weren't sustained by well-intentioned nonprofits or government programs. To say that young people growing up in West Baltimore lag far behind in Internet access and digital literacy is an understatement. Many students in public schools lacked Wi-Fi, basic educational materials, and even access to technology before the pandemic. During COVID, community leaders scrambled to provide laptops, but even that was insufficient. I thought about a student getting only the most basic education, who would then have to compete for access to higher education or in the job market with a student from Georgetown, where I teach. I once had a student who, in high school, had the opportunity to participate in a cooperative with Google, where she learned coding and about machine learning. She now competes for opportunities in the U.S. economy alongside students from Baltimore who barely had a chance at a decent basic education.

This situation in Baltimore reminded me of my own upbringing in New Orleans, where gross inequities persist. Like Baltimore, New Orleans bears the scars of redlining, colonialism, and historical injustices that divided the country in multiple ways, not just North vs. South, but across ethnic lines and beyond. Growing up, I had access to an excellent private school and the opportunity to engage with university campuses, even participating in university courses while still in high school.

It wasn't until I became an adult after Hurricane Katrina that I realized how different life was for students just two or three streets away from me. It had never been clear to me that housing and opportunity changed almost from block to block.

On one side, we had public housing projects where Title I schools often didn't have air conditioning in a city like New Orleans. These schools certainly didn't have computer labs, AP college prep classes, or access to sports and arts programs after school. Some students might not have even had an encyclopedia at home. Just beyond them were schools with soccer fields, scoreboards, and computer labs with Macintosh computers, which allowed students to begin learning how to participate in an increasingly digital society. Despite our physical proximity as children, the trajectories of our lives could not have been more different as we grew into adults. The opportunities we could access were vastly different because of something as simple as which side of the social borders we lived on. And this doesn't even begin to cover the challenges and inequities faced by the children who grew up in those housing projects.

The storm, which we name only as "Katrina," only unveiled and exacerbated those inequities. The city went through an entire remodeling of its public education system, trending toward charter schools, which displaced students and educators, further limiting access to resources. This period saw a doubling, tripling, or even quadrupling of gentrification and income inequality within the city. As rebuilding occurred, certain neighborhoods received top-tier infrastructure, while others were either left to their own devices or are still not fully recovered, with substandard infrastructure. This is a pattern I fear will repeat in the face of climate change, not just in the United States, but all over the world.

These examples highlight the long-term impacts of digital and climate inequities, particularly for youth. I focus on young people because they are emblematic of these systemic disparities and because these challenges deeply shape their futures. By focusing on youth, I also want to emphasize hope. This book advocates for scaling and accelerating community-driven solutions. Addressing digital equity is not just about education and economic opportunity. It is about bringing historically divested communities into the future, positioning them on par with connected, climate-resilient communities. Young leaders must be a part of that.

This should be a moral imperative. If only parts of the country are prepared to face the crises ahead, none of us will move forward. If only the global North is ready, the crisis will continue to create cascading challenges for the rest of the world. A holistic, asymmetrical approach is required to address the unique interactions between different contexts and locations. Only then can we achieve true justice, both digitally and environmentally. These historical injustices are not abstract; they persist in the lived experiences of communities like those in West Baltimore and New Orleans. The enduring effects of redlining, digital deserts, and environmental degradation are evident in these places, where disparities in access to resources are starkly visible.

As we stand at the crossroads of digital evolution, racing quickly from the early iterations of the Web to the generation of interactions that will encompass Web 3.0 and 4.0, the risk of perpetuating past injustices is considerable. This transition presents an unprecedented opportunity to redefine our social contract, embracing digital inclusion, human security, and proactive environmental stewardship. This book challenges us to envision a future that transcends historical inequities, advocating for a global renaissance of equity and justice.

The journey through this book is structured to engage, inform, and inspire. Starting with an examination of historical divestment and its contemporary manifestations, the narrative unfolds to explore the intersections of environmental, social, and digital justice. Through the prism of The Undivide Project, this discourse delves into the essence of holistic climate resilience in the digital era, confronting the stark realities of digital gentrification and the exploitative patterns of emerging technologies.

With a blend of empirical research and personal insights, this book navigates the complex interplay between the digital divide and climate vulnerability. Engaging with emergency managers through surveys, we unveil the critical linkage between digital equity and disaster resilience, highlighting the imperative for inclusive preparedness strategies.

The global dimension of these challenges is underscored in subsequent chapters, drawing parallels between the legacies of colonialism and contemporary climate vulnerabilities. Through a critical examination of policies and laws that have either hindered or fostered equity, the book builds a compelling case for a new paradigm of justice that integrates digital and environmental considerations into the fabric of global policy.

The final chapters of this book are a call to action, presenting a forward-looking perspective on how digital innovations can serve as catalysts for justice and equity. By showcasing community-led initiatives and advocating for comprehensive policy reforms, this book aims to galvanize a movement toward a more just and sustainable future. Through an analytical and quotidian narrative, "Climate & Code: Building Digital Justice in a Connected World" seeks to redefine the parameters of global justice, urging policymakers, academics, and activists alike to embrace a new ethos of inclusivity and sustainability. The book unfolds in a structured manner to delve deeper into the complexities and solutions surrounding climate and digital justice.

Washington, DC, USA Monica C. Sanders

Acknowledgments This book reflects the support, guidance, and generosity of many people.

I am deeply grateful to my partner in life and adventures, Edgar Diaz, for his unwavering support of all my endeavors. My family, especially Col. Edna Cummings, USA (Ret.), for her inspiration, leadership, and thoughtful advice. I thank the entire The Undivide Project team for their brilliance, commitment, and integrity in advancing community-centered work.

My sincere thanks to Anne Harris, Vicki Nolton, Susan Sanning, and everyone at Grinnell College and in Grinnell, Iowa, for a transformative experience. I am also grateful to my "sisters" in the Latin American Women's Network for Disaster Risk Reduction and WINDRR, as well as my colleagues in Disaster Researchers In Place, or the "DRIP," for their solidarity and intellectual companionship. Thank you to all of the communities and community organizations that worked with me over the years (Bayou Community Foundation, Rowdy Orb.it, Friends of Anacostia Park, the social justice team at ESRI, Ahora, University of Puerto Rico, and UTRGV colleagues, colleagues throughout the City of Baltimore, Kahea, Rooted, and so many more) and all of the emergency managers, public health, sustainability, and resilience professionals who helped us to understand. A special appreciation to Dr. Felicia Henry for her support in ensuring that I did not write something that did not do justice to the work.

Finally, I thank my colleagues in Panama, Centro Paula Souza, Georgetown University, and the University of London for their mentorship, generosity, and continued encouragement.

Competing Interests The author has no competing interests to declare that are relevant to the content of this manuscript.

About this Book

Climate & Code: Building Digital Justice in a Connected World examines how digital infrastructure, climate risk, and governance systems intersect to shape inequality, resilience, and opportunity. Drawing on interdisciplinary research, legal analysis, and community-based practice, the book advances a framework for digital justice that treats connectivity, data, and technology as essential public goods rather than neutral tools.

The book is intended for scholars, policymakers, legal practitioners, technologists, and institutional leaders working at the intersection of climate adaptation, digital governance, and social systems. It also provides a foundation for related convenings, workshops, and policy dialogues focused on aligning technology, law, and public accountability in an era of compound risk.

Contents

1

Climate and Code: Understanding the Intersection of Digital and Climate Justice

As this book goes to press, the global political landscape has shifted dramatically. Conflicts have arisen, and a number of countries in Europe and Central and South America have seen their leadership move away from democratic norms. In the United States, the 2024 elections ushered in a new federal administration that is reshaping regulatory priorities, funding streams, and public narratives around climate resilience, equity, and digital access. While not new, rollbacks in climate funding and the weakening of digital equity programs have become more urgent. These shifts reinforce the central claim of this book: that the existing social contract is rapidly failing, particularly for marginalized communities, and must be restructured to ensure resilience in the face of compounding digital and environmental risks.

This book draws from scholarly analysis, community experiences, and on-the-ground fieldwork, including work undertaken as a National Geographic Explorer. While personal reflection appears sparingly, it is used deliberately to contextualize professional experience and community insights. The aim is to reach both academic and practitioner audiences, including those building policy, writing code, or organizing on the ground, who need frameworks that are both rigorous and grounded in lived reality.

M. C. Sanders, *Climate and Code*, https://doi.org/10.1007/978-3-032-28032-9_1

The term "digital divide" has been used since the 1990s when the Internet began gaining prominence. Initially, it referred to the disparity in access to the Web and availability of hardware among different groups of people. Over the years, the concept of the digital divide has evolved to encompass not just access to the Internet but also the affordability and quality of that access. It also encompasses the availability of hardware, software, and the skills to take advantage of all the varying kinds of access. This is especially true of divested communities and communities of color.

The analysis that follows is grounded in original research conducted in communities across the United States and the Caribbean. This includes case studies from Baltimore, Maryland, and Adjuntas, Puerto Rico. I chose these locations not for their exceptionality, but because their conditions are emblematic of cascading digital and climate vulnerabilities. Community-level data collected through interviews, infrastructure mapping, and demographic analysis brings new evidence to the field of climate and digital justice.

Although progress has been made in addressing the digital divide, there are still significant gaps, particularly in terms of digital justice. The conversation must extend beyond access to include the nature of Internet infrastructure, as it is not equitably positioned in the United States and elsewhere. It should also consider issues of digital literacy, digital citizenship, access to public goods, and their intersection with economic mobility and climate change. If we acknowledge the existence of a social contract with marginalized communities, we must recognize the need to re-engineer that contract.

While this book often references U.S. policy and infrastructure, its scope is intentionally global. The structural dynamics that drive digital and climate injustice, colonial legacies, uneven development, privatized infrastructure, and policy inertia are not limited to geographic and political boundaries. Throughout this work, examples from Southeast Asia, the Caribbean, and Europe appear alongside U.S. cases to underscore the transnational and systemic nature of these challenges.

The Social Contract And How It Fails In This Era

Modern societies have long relied on an implicit social contract between governments, institutions, and citizens. Classical political theorists such as Thomas Hobbes, John Locke, and Jean-Jacques Rousseau (Lloyd, 2022; National Archives, n.d.; EBSCO Information Services, n.d.) described this contract as the arrangement through which individuals surrender certain freedoms in exchange for security, order, and the protection of shared rights. Over time, this contract expanded beyond political representation to include public goods such as transportation networks, electricity systems, public education, and environmental protections. These systems form the backbone of modern life because they enable citizens to participate in economic, social, and civic activity.

The concept of an informed and engaged citizenry lies at the center of this evolving social contract. Democratic societies rely on citizens who can access accurate information, communicate with institutions, and participate in decision-making processes that affect their communities. In the twenty-first century, digital infrastructure has become part of this foundational civic architecture. Reliable Internet connectivity, access to information networks, and digital literacy now shape whether individuals can fully participate in public life. As climate change increases the frequency and intensity of disasters, these systems also determine whether communities can access real-time warnings, emergency services, recovery resources, and the information necessary to make informed decisions. In this sense, digital infrastructure has evolved from a convenience into a form of essential public infrastructure.

Scholars such as Elinor Ostrom (Ostrom, 2000) have shown that communities are often most effective when they possess the knowledge and institutional capacity to manage shared resources and respond collectively to challenges. Similarly, work in climate governance highlights the importance of social capital, communication networks, and institutional trust in shaping how societies respond to environmental change. Communities with limited broadband access, low digital literacy, or fragile communication infrastructure may find themselves excluded from the very systems that enable resilience. This aspect of the so-called "digital divide" complicates the evolving social contract—perhaps points out its failures.

History of the Digital Divide

To fully understand the digital divide, it is crucial to examine its historical context and the cross-cutting impacts it has. One of the fundamental issues underlying the digital divide and the "contractual break" it represents is the inequitable distribution of infrastructure and its construction. The history of divestment, particularly in BIPOC (Black, Indigenous, and People of Color) communities, can be traced back to practices such as redlining and other forms of infrastructure inequities. In *Highway Robbery: Transportation Racism and New Routes to Equity*, Dr. Robert Bullard highlights how equal access to reliable and practical transportation is denied to many people, particularly the poor and people of color (Bullard et al., 2002). This lack of access to transportation networks restricts physical and social mobility, limiting opportunities for social, economic, and civic participation.

As previously noted, the concept of the "digital divide" emerged in the 1990s alongside the public adoption of the Internet. Over time, its meaning has evolved. Initially, it referred to the disparity in access to the Web and the availability of hardware, where some individuals and communities had access while others did not. This early understanding of the digital divide was primarily concerned with bridging the gap between those who had the means to connect to the Internet and those who did not.

In recent years, the concept of the digital divide has evolved to encompass broader dimensions of inequality. In 2022, during the Biden Administration's large broadband funding legislation, the term gained even more significance. It was no longer just about access to the Internet, but also about the affordability of that access and the availability of hardware and skills training, particularly in marginalized communities, including BIPOC communities (U.S. Congress, 2021).

The expansion of the digital divide discourse to include issues of affordability and hardware availability reflects the recognition that access alone is not sufficient. Simply having physical access to the Internet does not guarantee equal participation in the digital society. Without affordable access and the necessary hardware, individuals and communities are still left behind, unable to fully harness the opportunities and benefits offered by the digital age.

Brief History of the Internet and Digital Redlining

To understand the historical context of the digital divide, it is important to delve into the development of the Internet itself. The Internet emerged as a research project in the 1960s, primarily funded by the United States government. It was initially conceived as a decentralized network designed to withstand a nuclear attack, ensuring communication resilience for government actors (Computer History Museum, n.d.; Live Science, 2022). Over the years, the Internet has evolved into a global network connecting individuals, organizations, and governments worldwide.

In the early stages of the Internet's development, access was limited to academic and research institutions. The Internet gradually expanded to include commercial entities, and by the 1990s, it started to gain popularity among the general public. However, access to the Internet remained uneven, with disparities based on factors such as socioeconomic status, geographic location, and racial or ethnic background.

The concept of the digital divide gained prominence as policymakers and scholars began to recognize the implications of unequal access to the Internet. It became evident that those who were digitally excluded faced significant disadvantages in terms of education, employment, civic engagement, and social interactions. The term "digital divide" was used to highlight the need for interventions to bridge these gaps and ensure equitable access to the Internet and its associated benefits.

As the digital divide discourse progressed, it became increasingly clear that the divide was not solely a matter of individual access but was also influenced by structural and systemic factors. This realization led to the emergence of the concept of "digital redlining." Digital redlining refers to the discriminatory practices that result in certain communities, particularly low-income and marginalized neighborhoods, being systematically excluded from adequate Internet access and resources (CNET, 2021a, 2021b).

Historically, the term "redlining" originates from discriminatory practices in the housing market that began in the 1930s, where certain neighborhoods, primarily inhabited by minority populations, were deemed "high-risk" and denied access to housing loans and other resources. These practices perpetuated racial and economic segregation, creating lasting disparities in wealth, education, and opportunity.

Drawing on the historical legacy of redlining, one we will explore in detail in Chapter 3, digital redlining highlights how similar exclusionary practices continue in the digital realm. Redlined neighborhoods and communities with a history of divestment and systemic inequality often face limited affordable, reliable Internet access. Internet service providers may choose not to invest in infrastructure upgrades or expand their services in these areas, resulting in inadequate connectivity options and slower Internet speeds (CNET, 2021a, 2021b).

The same communities that were historically denied access to housing opportunities and other public goods now face challenges in accessing equitable Internet infrastructure. Internet infrastructure projects often build upon existing structures, which means that if a neighborhood is divided by highways or cut off from essential amenities, launching new Internet infrastructure projects in those areas perpetuates divestment.

Digital redlining deepens the existing inequalities and exacerbates the digital divide. It perpetuates a cycle of disadvantage, where communities already facing economic and social challenges are further marginalized in the digital landscape. The consequences of digital redlining are far-reaching, as they affect education, employment opportunities, access to critical services, and civic participation within these communities.

Recognizing the historical context and how the term "digital divide" has evolved to encompass issues of affordability and hardware availability, it becomes clear that addressing the digital divide requires a multifaceted approach. Bridging the divide necessitates not only expanding access to the Internet but also addressing structural inequalities, ensuring affordability, and challenging discriminatory practices such as digital redlining.

By doing so, society can move closer to achieving digital justice and equal participation in the digital age.

"Justice40" Funding, Program Design Flaws, and the Need for Federal Employee Education

The infrastructure bill passed early in the Biden Administration included funding for a variety of infrastructure projects, including efforts to improve Internet access. Policymakers built equity principles into these efforts through both legislative mandates and executive orders. One of the most prominent was Executive Order 14008, which launched the Justice40 Initiative and directed federal agencies to allocate forty percent of the benefits of applicable programs to communities marginalized by underinvestment and environmental injustice. At the time, this marked a potentially transformative shift in federal policy. The federal government acknowledged that systemic disinvestment had produced structural barriers to access, including broadband and climate resilience infrastructure (White House, n.d.; U.S. Congress, 2021).

That momentum did not hold. In 2025, the incoming Administration rescinded Executive Order 14008, effectively nullifying the Justice40 Initiative and its directives to federal agencies (Harvard Environmental & Energy Law Program, 2025). The rollback signaled a broader change in national priorities, weakening formal commitments to environmental and digital equity in federal policy-making.

The administration also ended or restructured related programs that aimed to expand Internet access and address digital inequality. The Affordable Connectivity Program (ACP) was one of the largest efforts under the Biden Administration to close the digital divide. Its design focused on expanding broadband access in unserved and underserved areas, particularly in rural and remote regions. The program provided monthly subsidies for broadband service and a one-time benefit for the purchase of devices such as laptops, tablets, and desktop computers (Federal Communications Commission, n.d.). It represented a step toward treating Internet access as a basic utility rather than a consumer luxury.

ACP aimed to directly support households in the greatest need. The idea was to remove cost barriers to connectivity, enabling more Americans to access education, healthcare, work, and emergency communications online. But the program faced challenges from the outset. It struggled with cumbersome registration systems, limited outreach, and documented cases of fraud (CBS News, 2023). More critically, it lacked long-term funding. In early 2025, Congress allowed the ACP to expire without renewal, abruptly cutting off support for millions of households (Congressional Research Service, 2025). The program, while flawed, had created temporary stability in access for many communities that once again found themselves disconnected.

A parallel effort, the Broadband Equity, Access, and Deployment (BEAD) Program, focused on infrastructure investment. Managed by the National Telecommunications and Information Administration, BEAD allocated $42.45 billion to support broadband planning, deployment, mapping, and adoption across all U.S. states, territories, and tribal lands (U.S. Congress, 2021). It was designed to promote infrastructure equity, meaning that investment would flow not only to areas with high commercial viability but also to those historically excluded from broadband buildout.

BEAD also included funding for training, digital literacy, and workforce development to ensure that newly connected communities had the tools to engage fully in the digital economy. The program acknowledged that closing the divide required not only cables and towers but also education and empowerment.

Like ACP, BEAD also became vulnerable to political shifts. In spring 2025, the U.S. Department of Commerce announced a policy shift under what it called the "Benefit Bargain" approach. This revision deprioritized equity benchmarks and streamlined requirements to reduce costs for private providers. The new guidance reframed BEAD as a tool for lowering regulatory burdens and expanding private sector investment, rather than for closing systemic access gaps (National Telecommunications and Information Administration, 2025). Even though the program was not formally canceled, its core commitment to digital justice was fundamentally weakened.

These reversals exposed deeper structural limitations. Even at their strongest, programs like Justice40, ACP, and BEAD only addressed federal policy. States, municipalities, and the private sector were not required to adopt similar equity frameworks, despite them controlling much of the country's broadband infrastructure. Communities' access to the Internet continued to depend on the financial feasibility of deploying broadband or satellite networks. In many cases, these decisions were governed by commercial interest rather than public need. The tools intended to guide equitable distribution of funds, the Federal Communications Commission's broadband maps, for example, were deeply flawed. The maps relied on inconsistent data and overestimated coverage in many areas. As a result, program administrators often misidentified which communities were unserved or underserved (CNET, 2021a, 2021b).

Beyond mapping, agencies lacked cohesive strategies for program rollout. A 2023 Government Accountability Office report documented the fragmented nature of federal broadband efforts and called for a national broadband strategy. The GAO urged federal agencies to adopt coordinated plans that integrated broadband expansion with other social and environmental goals, including education, climate adaptation, and economic mobility (Government Accountability Office, 2022).

Despite good intentions, the reality was more complicated. Effective program implementation required robust interagency coordination, sustained funding, and local input. As political priorities shifted, these conditions unraveled. The result was a return to pre-pandemic patterns of exclusion, with divested communities again left out of both infrastructure expansion and digital opportunity.

Sustainable solutions to the digital divide must move beyond short-term subsidy programs. They must account for geographic and economic barriers to deployment, particularly in regions where broadband is commercially unviable. This includes pursuing alternative models such as community-owned networks, mesh technologies, and low-earth orbit satellite systems. These approaches may not offer the same economies of scale as commercial broadband, but they can deliver tailored, place-based solutions that meet community needs.

Ultimately, federal investment alone will not solve the digital divide, especially if future administrations continue to dismantle the equity infrastructure that recent programs attempted to build. Long-term resilience requires not just infrastructure, but commitment: to inclusion, to coordination, and to recognizing access to information as a prerequisite for climate and economic justice.

Access to Infrastructure and Broadband Won't Solve Everything

One key aspect to consider is the limited scope of federal spending, which has primarily focused on the actions and initiatives of the federal government itself. This limitation remains especially significant given the ongoing challenges posed by population distribution, geography, and economic feasibility. These are all factors that continue to hinder progress in closing the digital divide. While past federal funding initiatives supported the construction of broadband infrastructure, there was never a consistent mandate or enduring incentive for legacy telecommunications companies to expand service or offer affordable rates in divested areas. Programs like the Affordable Connectivity Program (ACP) provided temporary subsidies, but they were not designed as long-term affordability solutions. Even in communities that benefited, connectivity remained uncertain once the funding ended (U.S. Congress, 2021).

Digital equity has always required more than federal investment. The responsibility for it extends beyond federal programs and involves the cooperation and coordination of states, cities, municipalities, and the private industry, which often owns the majority of the infrastructure. In the absence of enforceable standards or regulatory guidelines, legacy companies face few obligations to expand coverage or reduce costs for divested communities.

Infrastructure development itself often depends on whether broadband deployment is financially viable. This remains a core consideration for service providers and a structural barrier for underserved communities. While public funds may assist with initial construction, providers still bear the ongoing operational costs and often decline to serve areas where profit margins are too low.

Population distribution plays a significant role in these decisions. Sparsely populated or geographically isolated regions carry higher infrastructure costs per user, making them unattractive for commercial investment. Areas with sparse populations or located in remote regions face higher infrastructure costs per user, making it economically unviable for companies to invest in these areas. Then there are geographical challenges, such as rugged terrain or a lack of existing infrastructure, further compounding the difficulties in extending Internet services to these areas.

New entrants to the broadband market face distinct hurdles. Smaller or regional companies may struggle to increase service capacity quickly enough to meet demand. Scaling up infrastructure and expanding coverage to underserved areas can be a complex and costly process. Additionally, satellite companies, often seen as a potential solution for remote areas, continue to face limitations in speed, reliability, and affordability. Regulatory uncertainty also surrounds whether satellite and other emerging technologies qualify for future federal or state-level support (U.S. Congress, 2021). These limitations collectively restrict the range of viable options for expanding digital access.

As a result, while certain areas may gain infrastructure improvements, others will remain disconnected. These are often the same communities that have historically faced disinvestment, low-income, rural, Indigenous, and communities of color. In such places, access to the Internet is not a luxury but a necessity: essential for education, employment, healthcare, emergency response, and civic participation. Without sustained investment and a clear commitment to equity, the cycle of digital exclusion and reinforcement of structural inequalities is likely to persist. Ensuring that digital systems remain functional during disasters is, therefore, not only a technological challenge but also a governance issue.

Need for a Comprehensive Strategy and Evidence-Based Solutions

Addressing these continuing challenges requires a comprehensive approach. It involves collaboration between government entities, private sector actors, and community organizations to develop innovative solutions. This may include exploring alternative technologies, such as community-owned networks, fixed point networks, satellite-based connectivity, or wireless solutions, to reach underserved areas. Additionally, there is a need for policies and incentives that encourage competition, promote affordability, and prioritize underserved communities in broadband expansion efforts.

To overcome these challenges, collaboration between federal agencies, local governments, community organizations, and private sector partners is essential. By working together, these stakeholders can leverage their expertise and resources to develop effective strategies that bridge the digital divide and promote equitable broadband access. Regular evaluation, feedback, and adjustments should be incorporated into the programs to ensure that they adapt to the evolving needs and technologies, allowing all individuals and communities to reap the benefits of enhanced connectivity.

Need for Education and Awareness in Government—To Be Able to Provide the Same for Communities

The intersection of digital equity and climate justice has often been overlooked in workforce education for government agencies. Under previous federal leadership, there was a growing recognition that climate literacy and digital citizenship training were just as important for the staff designing government programs as for the communities they served. Executives and program officers administering equity-oriented initiatives were expected to build skills in data analysis, infrastructure planning, and community engagement. Program leaders and policy staff were to be trained to recognize and address biases in data collection, aggregation, and application (Sanders, 2023). Agencies like FEMA and the Environmental Protection Agency (EPA) had stated goals to embed equity into their missions and train their workforces accordingly.

Those initiatives have since stalled or reversed. In 2025, new political appointments and administrative priorities disrupted or eliminated many of these programs. A leaked transition memo detailed plans to abolish FEMA altogether or significantly weaken its authority, shifting responsibility to state and local governments and eliminating climate-specific mandates (Bloomberg CityLab, 2025). Similarly, the EPA's new administrator, Lee Zeldin, publicly committed the agency to "unleashing American greatness," signaling a sharp pivot away from climate and environmental justice initiatives (U.S. Environmental Protection Agency, 2025). These changes leave a vacuum in federal education and capacity-building around the connections between data, digital infrastructure, and climate resilience.

Before these rollbacks, the federal workforce development faced substantial challenges. Even at their peak, digital equity and climate justice training programs lacked unified standards. Federal program officers frequently cited the absence of baseline curricula or shared definitions for concepts like "climate literacy" or "data equity." FEMA, for example, had pledged to create a climate-literate workforce and integrate equity into all programs, but it lacked consistent frameworks to operationalize these goals. As a result, outcomes varied widely across departments and field offices.

At the EPA, similar issues existed in the management and application of data. Staff needed more training on how to use internally generated environmental data to inform local resilience strategies and make it accessible to communities. Those charged with delivering climate and digital justice outcomes often lacked the tools, guidance, or institutional support to do so effectively.

Disaster vulnerability, loss, and risk data remain critical for multiple areas of public service, from early warning systems and emergency coordination to insurance claims and post-disaster recovery. These data help communities adapt to climate impacts and advocate for resources. Despite the growing number of digital tools for data visualization and analysis, federal agencies continued to struggle with fragmented systems and inconsistent practices.

Many of these issues came from a lack of funding, mandates, or cross-agency infrastructure for consistent data collection and reporting. Only a few agencies are tasked with collecting disaster-related data, and state and local datasets for non-federally declared events often go unreported. The result is a patchwork of information systems, where critical metrics like housing damage, population displacement, and infrastructure loss are siloed and rarely interoperable. This problem is compounded by limited federal budgeting for disaster analytics and the absence of clear guidance on how to track public expenditures related to disaster response.

Several efforts had been underway to address these gaps, such as the development of national "all-hazards" datasets and the refinement of Incident Command System data standards. But these initiatives faced an uphill battle in scaling and integration. Work remained unfinished on incorporating cascading and long-term climate risks, including slow-moving disasters like heatwaves and chronic flooding, into national monitoring systems.

Existing datasets, such as those compiled by the National Interagency Fire Center, focused primarily on high-profile wildfire events. They largely excluded secondary impacts such as smoke exposure, public health effects, or damage to local economies. There was no standardized method to document damage to community assets, public infrastructure, or social vulnerabilities across events. Without this data, federal planners and taxpayers alike lacked a clear understanding of where resources were most needed or how effective interventions had been.

Federal agencies sit at the crossroads of digital and climate justice. Their ability to serve the public effectively depends on internal systems that support data-informed, community-centered decisions. The path forward must involve building institutional capacity: training staff to recognize the needs of historically marginalized populations, understand how infrastructure and connectivity impact resilience, and use data systems in ways that support transparency and trust. Standardizing and democratizing access to these tools and knowledge systems is not just a technical upgrade; it is a justice imperative.

Challenges Facing Communities—Where the Digital Divide and Climate Crisis Meet

As fraught as the challenges are with government actors, because of the aforementioned digital skills issues, access to and understanding of research and climate information at the community level is just as difficult. One reason is that the ability to share research insights also depends on Internet infrastructure and tools. Lack of Internet access or high-quality Internet access means the sensor networks upon which researchers often rely do not have the supportive infrastructure needed to collect critical climate and disaster data. Levin notes this in *Civic Internet of Things*: "Internet of Things (IoT) devices such as sensors can provide information that enables better infrastructure management and disaster planning. Data and broadband networks create opportunities to improve health care, education, public safety, job training, energy use, among other public services" (Levin, 2015, 297). Catherine Sandoval also warned of this dual challenge in the context of the 2019 Camp Fire in California and the 2020 COVID-19 Pandemic, "As the Internet's integration into society increases, the digital divide diminishes access to societal resources, including disaster aid, and exacerbates wildfire, flood, pandemic, and other risks. To mitigate climate change, and climate-induced disasters, protect access to social services and the economy, and safeguard democracy, it argues for digital inclusion strategies as a centerpiece of community-centered infrastructure regulation and disaster relief" (Sandoval & Lanthier, 2021).

These impacts present themselves in rapid-onset events like wildfires, protracted ones like the pandemic, and for some communities, create an everyday state of emergency. Consider Baltimore, Maryland, and the U.S. Territory of Puerto Rico. They are hyper-exposed to disasters, climate impacts, and poverty, but under-connected. I examined both in my research work and chose these two locations because they have the combination of the two highest child poverty rates and the lowest penetration of digital access in the country. My team and I examined the existence of high child poverty, digital dead zones, and disaster vulnerability to help articulate the acute impacts on children. This work was done in 2020 and 2021 during the COVID-19 pandemic. We found that in Baltimore, just over 40 percent of households had no wireline Internet access. That meant no cable, fiber optic, or other service. Parts of Baltimore are flood-prone, suffer from heat island impact, and face a variety of social mobility challenges.

As difficult as the context in this Maryland city may be, the U.S. territory of Puerto Rico has more concerning outcomes. While the island generally has Internet access, poorer communities are consistently left out of access and resilience-building resources. For example, in Adjuntas, one of the poorest cities in Puerto Rico, only 6 percent of households have any kind of access. In Puerto Rico, Baltimore, and generally, Indigenous, Black, and Latino children are at the highest risk for educational disruption because of a lack of access to the Internet. Over their lifetimes, these children will suffer from the immediate social and psychological impacts of educational disruptions, then later from the lack of economic opportunity resulting from an incomplete, substandard education (Sanders, 2021). All factors combined mean the population now and at least for another generation will struggle with access to public goods and information. This will hamper their ability to engage with important information about economic prospects and hazard mitigation that could help them build resilience in their communities.

Because these locations are disaster-prone and are so because of cascading risk, they and other areas need technological advancement and solutions support when it comes to emergency management. When we discuss the digital divide in terms of emergency management and climate risk, it is important to understand two things. First, disasters and emergencies are the rapid-onset symptoms of climate change that disproportionately impact divested communities. Second, when discussing emergencies, we are mainly discussing communications decisions over research, monitoring, and data aggregation. One of the more dramatic examples of how the digital divide impacts communities in this scenario is the case of the 2020 Western Wildfires. As wildfires raged in Jackson County, Oregon, in September 2020, local officials declined to activate their Emergency Alert System, leaving television and radio programs uninterrupted and sending emergency alerts only to residents who had signed up for an online notification system (Tankersley, 2020, citing Sandoval). Poor decisions about technology deployment left thousands with little to no information, dependent on neighbors and glimpses of the orange sky for fire warnings. At the low end, people who did not sign up for alerts did not receive critical information, and people without Internet access or dependent upon data, meaning they restrict phone usage because of the cost of data plans, another digital divide impact, never or rarely received critical information. The result is more than risk; it is immediate danger of loss of property or life (Sandoval & Lanthier, 2021).

Considering the historical context of divestment and the understanding of the information divide, we can easily guess what kinds of communities bear the brunt of digital divestment and climate change. In the 2020 United States Census, we learned that 60 percent of African Americans live in just ten states, the majority of which are in the South and along the Gulf and Mid-Atlantic coasts. These are all known to be areas at outsized risk of sea-level rise, heat impacts, and severe weather. Globally, we know that close to 300 million people living in the hottest and most flood-vulnerable areas of Southeast Asia have no access or very low-quality access to the Internet. When these groups are interacting online, it is often via mobile phones and data plans that can be expensive.

Further compounding the challenges within the same groups of people is that another impact of being divested is that these populations are also under-skilled or lack the digital literacy to be able to interact with these platforms in a meaningful way. The issue is widespread. For example, a quarter of adults in Europe, mostly Central and Eastern Europe, have low to no basic digital literacy skills. According to the National Skills Coalition, about a third of adults in the United States lack basic digital literacy (National Skills Coalition, 2022). Basic digital literacy is defined as having the skills you need to live, learn, and work in a society where communication and access to information are increasingly through digital technologies like Internet platforms, social media, and mobile devices (Western Sydney University, 2020).

This lack of basic digital literacy, not just a lack of access, but a lack of ability to interpret digitally distributed information, supports the idea that the digital divide is also a climate risk multiplier. John Roome of the World Bank wrote in a 2016 World Bank report, "In many developing countries, more families now own a mobile phone than have access to electricity or clean water. No major technology has reached more people in such a short time, but unfortunately, there is still a significant digital divide between the poor and the wealthy parts of the global population when it comes to Internet access. Perhaps not surprisingly, the same digital divide has an impact on the ability of developing countries to deal with the impact of climate change. Climate risk management requires data and knowledge. Connectivity is therefore a fundamental part of the equation for protecting poor people from climate change" (Roome, 2016).

The idea of public or in-home Wi-Fi is elusive for many people in extremely high-risk climate areas. Communities without high-quality access to information or the skill set to interpret and make decisions about climate and disaster information cannot participate in mitigation and adaptation activities. Access to accurate and timely climate and disaster information is crucial for informed decision-making and effective response strategies. Without the ability to interpret and utilize digital resources, these communities are further marginalized in their efforts to address climate risks and adapt to changing environmental conditions.

Adding to the complexity of the problem is that researchers rely on sensor networks to obtain critical climate data. Those sensors need a connection to the Internet, mainly Wi-Fi, to produce consistent data. Similarly, emergency management and humanitarian aid organizations have developed a broad portfolio of climate risk tools, open datasets, mapping projects, and social media flood tagging programs, all of which rely on the Internet. When there is no connection or a poor-quality connection, information does not consistently flow into or out of the communities with the highest climate risk.

In addition to climate risk and tenuous access to the Internet and digitized public goods, people in least-developed countries and divested communities in the United States also face financial inclusion problems, as validated in the *Shock Waves* report which focuses on poverty and climate change (Hallegatte et al., 2016). Poorer communities also broadly lack access to capital and financial tools, which are also heavily Internet-access reliant, that can help them with mitigation and adaptation activities.

Those activities can also be economic drivers and help with upskilling. Enhancing digital skills and providing climate adaptation training can empower communities to overcome these challenges and leverage the benefits of concepts like regenerative finance and artificial intelligence tools. Emerging fields like regenerative finance focus on creating financial systems that promote social and environmental well-being, aiming to generate positive impacts beyond mere financial returns. By acquiring digital skills and literacy, communities can access regenerative finance platforms and similar tools, enabling them to explore sustainable investment options, secure funding for climate action initiatives, and actively participate in the development of regenerative economies.

Empowered with the necessary digital literacy, individuals within these communities can actively participate in decision-making processes, advocate for their needs, and access resources supporting climate resilience and adaptation. This not only enhances their ability to navigate the digital landscape but also amplifies their capacity to contribute meaningfully to climate action, fostering greater equity and sustainability in the face of climate challenges.

Creating New Thought Processes, Frameworks, and Opportunities

Knowing that the cascading and often overwhelming impacts of how the digital divide interplays with climate risk have been brought forth for examination, global society has a moral imperative to address these impacts. The design of interventions must be interdisciplinary, bringing together researchers, policy professionals, and legal experts. It also has to take place at operational, legal, and policy levels and confront how we view civil liberties. It must also be as intersectional as those interventions that focus on equally far-reaching challenges like equity.

If Metcalfe's law, which states that the value of a telecommunications network is proportional to the square of the number of connected users of the system (n2), holds, then we can extend this concept to a capitalist system (Hendler & Golbeck, n.d.). In a capitalist system, increased access to capital leads to higher production and value across economies.

This understanding challenges the notion of financial disinclusion, which refers to uneven access to financial resources. Similarly, it raises questions about the persistence of climate, environmental, and digital injustices, which hinder equitable access to opportunities and resources.

Considering this, we must contemplate how to create holistic resilience while challenging imperfect systems. If the value of a network increases with more users, should we not strive for an inclusive network that benefits all? The same applies to our economic system. Should access to capital not be available to all, fostering greater productivity, value creation, and ecosystem remediation?

The reality is that financial, climate, environmental, and digital injustices continue to persist. The social contract that once centered on physical infrastructure and public institutions must now extend to the digital networks that increasingly organize how societies communicate, respond to risk, and recover from disruption.

A new social contract would envision a more inclusive and equitable digital landscape, focusing on uplifting divested communities under renewed terms. This prompts reflection on how to forge a path toward holistic resilience and address these systemic issues. The work would have to include a path toward collaborative governance and a commitment to dismantling the barriers that perpetuate inequality. It would also involve building systems and structures that prioritize inclusion, diversity, and sustainability. By reimagining the social contract, it is possible to strive for a future where digital, financial, climate, and community systems can thrive.

References

Bloomberg CityLab. (2025, June 18). *New memo details how Trump plans to abolish FEMA.* https://www.bloomberg.com/news/newsletters/2025-06-18/new-memo-details-how-trump-plans-to-abolish-fema-citylab-daily

Bullard, R. D., Johnson, G. S., & Torres, A. O. (2002). *Highway robbery: Transportation racism and new routes to equity.* South End Press.

CBS News. (2023, February 6). *Federal program that helps millions pay for internet access is plagued with fraud and complaints*. https://www.cbsnews.com/texas/news/i-team-internet-access-fraud-complaints/

CNET. (2021, February 19). *Millions of Americans can't get broadband because of a faulty FCC map. There's a fix*. https://www.cnet.com/home/internet/features/millions-of-americans-cant-get-broadband-because-of-a-faulty-fcc-map-theres-a-fix/

CNET. (2021, June 28). *The broadband gap's dirty secret: Redlining still exists in digital form*. https://www.cnet.com/home/internet/features/the-broadband-gaps-dirty-secret-redlining-still-exists-in-digital-form/

Computer History Museum. (n.d.). *Internet history 1962–1992*. https://www.computerhistory.org/internethistory/

Congressional Research Service. (2025). *The Affordable Connectivity Program: Background and status* (IF12637). https://www.congress.gov/crs-product/IF12637

EBSCO Information Services. (n.d.). *Social contract: Jean-Jacques Rousseau*. EBSCO Research Starters. https://www.ebsco.com/research-starters/literature-and-writing/social-contract-jean-jacques-rousseau

Federal Communications Commission. (n.d.). *Affordable Connectivity Program*. https://www.fcc.gov/acp

Government Accountability Office. (2022). *Broadband: National strategy needed to guide federal efforts to reduce digital divide* (GAO-22-104611). https://www.gao.gov/products/gao-22-104611

Hallegatte, S., Bangalore, M., Bonzanigo, L., Fay, M., Kane, T., Narloch, U., Rozenberg, J., Treguer, D., & Vogt-Schilb, A. (2016). *Shock waves: Managing the impacts of climate change on poverty*. World Bank. https://openknowledge.worldbank.org/handle/10986/22787

Harvard Environmental and Energy Law Program. (2025). *Rollback: Trump rescinded Biden's Executive Order 14008 that established the Justice40 Initiative*. https://eelp.law.harvard.edu/tracker/rollback-trump-rescinded-bidens-executive-order-14008-that-established-justice40-initiative/

Hendler, J., & Golbeck, J. (n.d.). *Metcalfe's law, Web 2.0, and the Semantic Web*. University of Maryland. https://www.cs.umd.edu/~golbeck/downloads/Web20-SW-JWS-webVersion.pdf

Live Science. (2022, April 8). *Internet history timeline: ARPANET to the World Wide Web*. https://www.livescience.com/20727-internet-history.html

Lloyd, S. A. (2022). *Hobbes's moral and political philosophy*. Stanford Encyclopedia of Philosophy. https://plato.stanford.edu/entries/hobbes-moral/

National Archives. (n.d.). *Founders Online: Chapter 4, Section 1* [Historical document collection]. University of Chicago Press/National Archives. https://press-pubs.uchicago.edu/founders/documents/v1ch4s1.html

National Skills Coalition. (2022). *The new landscape of digital literacy*. https://nationalskillscoalition.org/resource/publications/the-new-landscape-of-digital-literacy/

National Telecommunications and Information Administration. (2025). *Trump Administration announces "Benefit Bargain" BEAD program*. https://www.ntia.gov/press-release/2025/trump-administration-announces-benefit-bargain-bead-program-removes-regulatory-burdens-lowers-costs

Ostrom, E. (2000). Collective action and the evolution of social norms. *Journal of Economic Perspectives, 14*(3), 137–158. https://doi.org/10.1257/jep.14.3.137

Roome, J. (2016, January 14). *The digital divide: A challenge to overcome in tackling climate change*. World Bank Blogs. https://blogs.worldbank.org/climatechange/digital-divide-challenge-overcome-tackling-climate-change

Sanders, M. C. (2021). *The confluence of poverty, the digital divide and disaster vulnerability on children: Baltimore and Puerto Rico as case studies. Academia Letters*. https://doi.org/10.20935/AL1478

Sanders, M. C. (2023). *Using a digital justice framework to improve disaster preparation and response*. Federation of American Scientists. https://fas.org/publication/using-a-digital-justice-framework-to-improve-disaster-preparation-and-response/

Sandoval, C. J. K., & Lanthier, P. (2021). Connect the whole community: Leadership gaps drive the digital divide and fuel disaster and social vulnerabilities. *SSRN Electronic Journal*. https://doi.org/10.2139/ssrn.3766610

U.S. Congress. (2021). *Infrastructure Investment and Jobs Act* (H.R. 3684, 117th Congress). https://www.congress.gov/bill/117th-congress/house-bill/3684/text

U.S. Environmental Protection Agency. (2025). *What they are saying: EPA Administrator Lee Zeldin to "unleash American greatness"*. https://www.epa.gov/newsreleases/what-they-are-saying-epa-administrator-lee-zeldin-unleash-american-greatness-head-epa

White House. (n.d.). *Justice40: A whole-of-government initiative*. https://www.whitehouse.gov/environmentaljustice/justice40/

Western Sydney University. (2020, November 26). *What is digital literacy?* https://www.westernsydney.edu.au/studysmart/home/study_skills_guides/digital_literacy/what_is_digital_literacy

2

Holistic Climate Resilience in the Digital Era

One of the ways in which the deficiencies in the current social contract have been shown is through the uneven distribution of climate risk and digital infrastructure across communities. Envisioning and then designing a new social contract for climate resilience in the digital era is to understand the concept of resilience. Resilience is often misunderstood, especially in the digital age, where information deficits and digital inequities challenge communities' ability to "build back better." In reviewing traditional resilience-building scholarship, there was plenty of emphasis on the capacity to "bounce back" from disruptions, focusing on returning to pre-disaster conditions. This approach is rooted in psychological, ecological, and socio-ecological frameworks that define resilience as the ability to absorb, adapt, and recover from adverse events (Chuang et al., 2018). Key concepts include robustness, recovery, and resourcefulness, highlighting individual and community capacities to withstand shocks, such as natural disasters, economic downturns, or social conflicts (Serfilippi & Ramnath, 2018).

The issue is that this overlooks systemic issues such as unequal access to digital resources. By challenging traditional models, we emphasize the need for approaches prioritizing equity, access, and empowerment,

M. C. Sanders, *Climate and Code*, https://doi.org/10.1007/978-3-032-28032-9_2

aiming to address the root causes of inequality and digital exclusion. There are opportunities to expose the everyday struggles of those impacted by digital disparities, revealing the real-world consequences of unequal access to technology and information. This perspective encourages a shift toward a more inclusive and holistic vision of resilience in the digital age, inspiring transformative change through integrated and equitable strategies.

Overview of Literature and Models

The ecological concept of resilience, as developed by Holling, characterized resilience as the capacity of ecosystems to absorb change and persist. This model was later adapted to social systems, emphasizing the need to build community capacity by strengthening local institutions, enhancing social capital, and promoting adaptive management practices (Horney et al., 2018).

Despite having a foundational role in scholarship and many practical applications, traditional resilience frameworks have been criticized for their limited attention to systemic inequities, such as socioeconomic disparities and unequal access to resources. Scholars argue that these models often overlook the dynamic nature of resilience and fail to address the broader social and technological factors influencing communities' adaptive capacities, particularly in marginalized populations (Cai et al., 2018). This gap has prompted the evolution of more integrative approaches that consider resilience within complex social, economic, and technological systems.

Resilience and Digital Access

Recent scholarship expands traditional resilience models by incorporating the role of digital access, acknowledging that information and communication technologies (ICTs) significantly impact resilience-building efforts. Digital access, which encompasses Internet connectivity, digital literacy, and access to data, has emerged as a crucial element in enhancing community resilience, particularly in the context of disaster preparedness, response, and recovery (Chuang et al., 2018). Digital inequities create significant barriers for marginalized communities, limiting their ability to engage with critical information during crises and participate in digital economies and hindering adaptive capacity.

Digital exclusion prevents communities from fully benefiting from early warning systems, online education, remote work opportunities, and other digital tools that can enhance resilience (Cai et al., 2018). This highlights the need for equipping communities with the ability to access and interpret climate data, engage with emerging technologies, and develop localized solutions to climate challenges. The shift from traditional resilience models to those that incorporate digital access reflects a broader recognition of technology's role in fostering more inclusive and adaptable communities (Serfilippi & Ramnath, 2018).

Resilience and Nature-Based Solutions

Another modern approach to resilience emphasizes the integration of nature-based solutions (NBS). These are strategies that use natural systems and processes to address environmental challenges, such as climate change, biodiversity loss, and ecosystem degradation. NBS, such as wetland restoration, green infrastructure, and sustainable land management, provide a cost-effective and sustainable approach to enhancing resilience. However, these solutions can be significantly bolstered by the integration of advanced technologies. The use of technology in NBS allows for better monitoring, planning, and management, making them more effective in addressing climate challenges, such as restoring wetlands to absorb floodwaters, planting urban trees to reduce heat island effects, and preserving coastal mangroves for storm protection (Bertilsson et al., 2018).

Critics of NBS point out several challenges. One major criticism is the lack of public understanding and clear definitions, which can hinder effective implementation. Communities may resist NBS due to fears of change, with concerns about how these interventions will alter their environment. Additionally, the equity of implementation remains a concern, as NBS must address the needs of marginalized communities to ensure broader social resilience (Nelson et al., 2020). These criteria underscore the importance of sustained engagement and inclusive decision-making processes, ensuring that all voices are represented and heard in the planning, implementation, and execution of nature-based solutions.

The intersection of technology and NBS has also gained traction, with digital tools being used to monitor environmental changes, manage natural resources, and engage communities in resilience planning. For example, sensor networks, remote sensing, and geospatial analysis enable real-time data collection, helping communities to anticipate and respond to environmental risks more effectively (Moghadas et al., 2019). Public–private partnerships often drive these innovations, bringing together the expertise and resources needed to scale up technology-supported NBS.

Innovative solutions in AgTech and green waterfront development offer compelling examples of how nature-based strategies can be successfully integrated with technological advancements. AgTech innovations, such as precision farming and data-driven regenerative agriculture, have proven to be effective in enhancing ecosystem health while improving productivity and supporting local economies. These technologies allow farmers to tailor their practices to the unique needs of the land, ultimately promoting both environmental and economic sustainability. Similarly, green waterfront developments, like the Cheonggyecheon Stream Restoration in Seoul, South Korea, demonstrate how nature-based strategies can revitalize urban areas, enhance social cohesion, and provide environmental benefits. These projects, which integrate natural infrastructure into cities, improve climate resilience while promoting public health and recreation opportunities (Ajiboye et al., 2020). The combined use of digital tools and NBS represents an approach that addresses immediate needs while promoting long-term sustainability (Serfilippi & Ramnath, 2018).

Comparative Analysis

Comparing traditional resilience models with modern approaches reveals a significant shift toward more holistic and systemic strategies. Traditional resilience-building focuses primarily on the ability to recover from specific shocks, often neglecting the underlying social, economic, and technological dimensions that shape resilience (Chuang et al., 2018). In contrast, contemporary frameworks emphasize the importance of digital inclusion and the integration of nature-based solutions, recognizing that resilience is not merely about recovery but also about adaptation, transformation, and equity.

The inclusion of digital access and NBS in resilience planning reflects a growing awareness of the interconnectedness of social, environmental, and technological factors. Digital access enables communities to better navigate and respond to crises, while NBS offers sustainable alternatives to conventional infrastructure that enhance both ecological and social resilience. Together, these approaches provide a more comprehensive understanding of resilience, one that addresses the root causes of vulnerability and prioritizes the needs of marginalized groups (Bertilsson et al., 2018; Moghadas et al., 2019).

The Work of Creating Equitable Digital Access and Resilience

Equitable digital access is a cornerstone of holistic resilience, but achieving it requires more than expanding physical infrastructure. The digital divide creates profound disparities, limiting individuals and communities in accessing critical services, education, and economic opportunities. As the Aspen Institute's "Roads to Resilience" initiative emphasizes, rural communities face unique barriers. Those include issues like sparse infrastructure, higher costs, and policy gaps that demand tailored solutions for digital equity (Aspen Institute, 2024). Part of holistic digital equity work is about tackling gaps in digital skills and literacy. For many, digital training focuses on basic tasks like navigating the Internet, using email, and identifying scams or misinformation.

These are essential, but not enough. Communities also need access to advanced skills that support the use of learning management systems, online platforms, and digital certifications. These skills open doors to emerging spaces. When digital literacy programs are accessible and well-designed, they can help marginalized communities use connectivity to pursue education, secure employment, and build wealth. All of which are key steps toward closing the digital divide and building resilience.

Despite this promise, digital exclusion continues to follow systemic lines. Dorries and Rhoads (2021) describe "information redlining," where underserved neighborhoods are denied quality digital services. This mirrors the history of redlining in housing and reinforces racial and economic inequity. These same communities often face heightened climate risks, creating overlapping challenges that block adaptation and recovery. While digital upskilling and climate literacy could help build resilience, existing job programs frequently fall short. Historic bias continues to shape how these programs are designed and delivered, often excluding rural and Black communities from more advanced training. Instead of being guided toward roles in emerging sectors, many are funneled into manual labor under the banner of "green jobs." Few are offered education in areas like coding, blockchain, or data science. Even fewer receive training that reflects their lived climate risks or equips them to participate in the research shaping future resilience strategies and economic opportunities.

The impacts of the digital divide are also deeply intersectional. Women in many regions face lower rates of digital access due to cultural norms, caregiving responsibilities, or financial constraints. People with disabilities often encounter inaccessible platforms and devices, excluding them from digital resilience tools. Indigenous communities risk being overlooked in mainstream resilience planning despite possessing valuable knowledge about local ecosystems. Addressing these overlapping inequities requires digital inclusion strategies that are gender-responsive, disability-inclusive, and culturally respectful (UNESCO, 2021; World Bank, 2020).

When they existed, federal climate initiatives like the Climate Corps and the National Climate Resilience Framework fell short in integrating technology-focused green jobs. These programs tend to emphasize entry-level or manual labor roles, overlooking the opportunity to equip communities with skills in fields like artificial intelligence, data analysis, or blockchain. These emerging sectors are increasingly essential for both climate resilience and economic mobility. As noted in the Aspen Institute's policy dialogues, incorporating technology into workforce programs, especially in rural and marginalized communities, can help close equity gaps and support long-term local prosperity (Aspen Institute, 2024).

BREAKOUT BOX: The Future of Federal Workforce Preparedness

As climate risks rise, federal agencies are increasingly recognizing that digital equity is a resilience issue. Recent discussions point toward integrating digital literacy into federal workforce preparedness programs. These efforts aim to ensure public servants—from emergency managers to social service providers—can use emerging technologies to support crisis response and recovery (Sanders 2023a).

Despite this potential, U.S. government efforts have missed key opportunities to promote innovation and inclusion. Technology experts and rural advocates have pointed out that investments in AI, machine learning, and related tools were either underutilized or targeted at limited audiences. Many of the newly created roles were designed for youth programs or required advanced degrees.

Meanwhile, the positions offered in rural and historically excluded areas were more often described as "construction and manufacturing" jobs (Sanders, 2023a, 2023b). This mismatch limits the broader potential of green workforce development to serve as a bridge to digital equity and climate adaptation.

The digital divide reflects broader systemic barriers to education, employment, and essential services. It also impairs access to life-saving information during emergencies, weakening local response and recovery systems. As highlighted by LINK Health and partners working in the

Mississippi Delta, rural communities are already navigating overlapping health, climate, and infrastructure challenges. Digital access must be embedded in broader strategies for community development and resilience (LINK Health, 2024).

The *Roads to Resilience* summit also emphasized that broadband and digital infrastructure should be treated as essential infrastructure, alongside roads, clean water, and energy. Rural leaders stressed that when broadband systems are built with community input and maintained with equity in mind, they can improve education, disaster preparedness, and economic development (Aspen Institute, 2024). For these communities, bridging the digital divide is not only about connectivity. It is about ensuring full participation in climate solutions, public decision-making, and future economic systems. Digital inclusion, then, is a social, environmental, and economic imperative that must be addressed through community-centered strategies.

The Digital Divide and Climate Resilience

Climate change poses significant risks to communities worldwide, including increased frequency and severity of extreme weather events such as floods, droughts, and heat waves. For vulnerable populations, the digital divide amplifies these risks by limiting access to early warnings, emergency information, and resources that are crucial to building resiliency. In marginalized urban areas, digital inequities often overlap with environmental vulnerabilities. The lack of access to digital tools and platforms restricts community engagement in resilience planning and limits the ability to participate in broader climate adaptation initiatives (Centering at the Margins: Critical Community Resilience Praxis). The digital divide not only prevents access to information but also excludes communities from the decision-making processes that shape their future resilience.

In South Asia, communities facing severe climate impacts such as rising temperatures and changing precipitation patterns are often also grappling with limited access to digital resources. The absence of timely information exacerbates vulnerabilities in communities already struggling with inadequate infrastructure and financial constraints (Climate Change and Community Resilience: Insights from South Asia). Access to digital platforms for information sharing, weather forecasting, and disaster preparedness can significantly enhance community resilience, yet these resources remain out of reach for many.

Similarly, innovative solutions are emerging across Sub-Saharan Africa. For instance, Kenya's Ushahidi platform enables communities to crowd-source real-time reports of flooding, drought, or infrastructure damage, empowering local actors to prioritize resources and advocate for government response (Okolloh, 2009). In Latin America, digital mapping initiatives in Brazil and Colombia combine indigenous knowledge with GIS tools to identify climate vulnerabilities in rainforest communities, ensuring resilience strategies are culturally relevant and ecologically effective (Rodrigues et al., 2020). These examples highlight how digital technologies can support resilience while being adapted to diverse local contexts.

Ensuring that digital literacy training and support programs are available can empower individuals in marginalized communities to leverage broadband connectivity for education, employment, and wealth creation. Access to these opportunities should be a core objective of efforts to reduce the digital divide. Unfortunately, historical inequities influence the administration of digital upskilling and job training programs. Often, digital literacy programs perpetuate biases against residents of rural communities and can be anti-Black. Many programs focus only on rudimentary skills without providing pathways to advanced education or high-paying jobs, thus reinforcing existing socioeconomic disparities.

The same is true across industries and programs; members of divested communities are frequently offered "green jobs," but these roles often focus on manual labor rather than positions in higher-wage, technology-driven sectors. Digital literacy training frequently does not include valuable skills such as coding, blockchain, or data science. These topics are more often reserved for urban and non-Black residents. Moreover, these training programs rarely address the context of climate risk, failing to equip participants with the skills needed to understand, manage, or mitigate their specific environmental vulnerabilities.

Another glaring gap in current digital upskilling efforts is the lack of training on how to engage with and interpret research relevant to both digital and climate challenges. This omission leaves communities without the critical insights needed to make informed decisions about their resilience strategies. Effective digital skills training should encompass a broader range of capabilities, including data analysis, environmental monitoring, and the use of digital tools to advocate for community needs in the face of climate change.

Digital Tools as Enablers of Community Development and Resilience

Despite the challenges posed by digital inequities and access for divested communities, technology offers significant potential to support development and resilience-building efforts. Digital tools can provide critical information, facilitate community engagement, and enable residents to participate actively in resilience planning. For instance, mobile apps, social media platforms, and online forums can enhance information sharing, allowing communities to stay informed about local climate risks and collaborate on adaptive strategies.

Digital literacy programs that empower individuals to use these tools effectively can transform how communities engage with resilience initiatives. Beyond basic skills, training that includes how to access online certifications, navigate learning platforms, and understand digital finance tools can open doors to new economic opportunities. For example, learning to use platforms like LinkedIn Learning or Coursera can connect residents with courses and certifications in fields such as data analytics, environmental management, and digital marketing, enabling them to access higher-paying jobs and diversify their skill sets.

Advanced digital skills training in coding, blockchain, and data science can provide divested communities with access to rapidly growing sectors of the digital economy. By equipping individuals with these skills, digital literacy programs can help break the cycle of poverty and exclusion that often characterizes communities on the wrong side of the digital divide. Importantly, such training should also include elements of climate literacy, helping participants understand how digital tools can be leveraged to address local environmental challenges.

Despite persistent digital inequities, technology holds the potential to support both community development and resilience. Digital tools can provide access to critical information, foster civic engagement, and enable communities to design solutions that respond to their specific challenges. Examples range from mobile apps that deliver real-time weather alerts to platforms that support participatory budgeting and collaborative planning for infrastructure needs. These tools are reshaping how communities prepare for and respond to climate risks.

Globally, cities are exploring how digital infrastructure and nature-based solutions can work together to promote resilience. Metz (2023) outlines a vision of sustainable urban spaces that integrate smart systems with green design, embedding resilience into daily life. Similarly, Regnier (2023) highlights how cities like Rotterdam and Seoul have combined revitalized waterways with digital monitoring systems to manage flood risks while enhancing public space and access. These cases illustrate how technology and nature-based approaches can converge to support both ecological and social outcomes.

Still, these benefits are not shared equally. Digital redlining continues to deny many low-income and marginalized neighborhoods high-quality broadband and related services (Dorries & Rhoads, 2021). This persistent gap limits access to timely information, reduces opportunities for participation, and weakens the overall capacity for resilience planning. Schindler and Wolf (2020) have noted the growing role of advanced tools such as big data analytics, remote sensing, and machine learning in climate adaptation, particularly for optimizing nature-based solutions. Without equitable access and the necessary skills to engage with these tools, however, the risk is that technology will further entrench existing disparities.

Digital literacy efforts need to move beyond basic instruction to include training in platforms and tools that support both community knowledge and economic opportunity. Skills in GIS mapping, environmental data analysis, and the use of digital advocacy tools can allow residents to participate in resilience-building work directly. Platforms like LinkedIn Learning and Coursera can also open pathways to professional certifications in fields such as data science, environmental management, and digital marketing, offering opportunities for economic mobility alongside increased climate awareness.

In addition, emerging Web 3.0 technologies, such as blockchain and decentralized applications (dApps), are being explored as tools for community-driven planning. These tools allow for transparent data collection, decentralized funding, and improved accountability in public engagement processes. For example, pilot programs in Europe are using blockchain to record community input on green infrastructure projects, increasing transparency and trust in how decisions are made (Schindler & Wolf, 2020).

To realize the full potential of these tools, however, communities must be equipped with both the infrastructure and the knowledge to use them. Overcoming these barriers requires sustained investment in broadband access, community technology hubs, and inclusive education programs that reflect the lived experiences and leadership of frontline communities. As highlighted in the Aspen Rural dialogues, digital capacity building is not simply a technical challenge; it is central to advancing social equity and climate resilience.

When accessible and integrated with community goals, digital technologies can shift resilience efforts from reactive recovery toward proactive planning. They provide a foundation for communities to anticipate risks, share knowledge, and shape their futures in the face of climate change.

Expanding Digital Literacy and Climate Literacy

Expanding digital and climate literacy together empowers residents to make informed decisions about resource management, disaster preparedness, and resilience planning. Many existing programs miss this integration, focusing narrowly on basic digital skills and neglecting to connect these skills to practical applications in managing climate risks. Effective programs should address these dual needs, ensuring that residents not only acquire valuable digital skills but also understand how to apply them in the context of their climate risks. This intersection of digital and climate literacy is crucial for enabling communities to make informed decisions about resource management, disaster preparedness, and resilience planning.

Similar to programming for digital skills, climate literacy programs often focus narrowly on basic concepts, neglecting to equip participants with the knowledge needed to engage with complex climate issues. This shortfall limits the potential of community-focused programs to contribute to holistic resilience. For example, residents who understand how to use GIS tools could participate in mapping local flood risks or identifying areas for green infrastructure, yet these skills are rarely included in standard digital training curricula.

Digital upskilling and climate literacy programs must also address the broader systemic barriers that prevent marginalized groups from accessing advanced training opportunities. This includes confronting biases that limit access to high-skill training for rural, Black, and Indigenous communities. For instance, while urban residents may have access to coding boot camps, data science workshops, and eco-entrepreneurship accelerators, similar opportunities are often unavailable or inaccessible to residents in rural or economically disadvantaged areas. Addressing these disparities requires targeted investments in digital education and infrastructure, as well as efforts to create inclusive learning environments that support diverse learners.

Using Technology to Accelerate Solutions in Communities

Technology can play a transformative role in accelerating resilience solutions within communities, but only if access to these tools is equitable and inclusive. From local governments using data analytics to optimize emergency response efforts to community groups leveraging social media for organizing and advocacy, digital tools offer powerful means to enhance resilience at the grassroots level. To fully harness these technologies, communities must first overcome the digital barriers that limit participation.

There are rural and low-income communities that are leveraging digital platforms to build local capacity and enhance their resilience to climate impacts. In South Asia, traditional knowledge is being combined with modern technologies like water management systems and sustainable agriculture practices, showcasing how technology can complement community-led resilience efforts (Climate Change and Community Resilience: Insights from South Asia). Similarly, digital platforms that facilitate remote learning, professional networking, and virtual collaboration are providing new opportunities for skills development and economic participation in marginalized communities.

Scaling these successes requires a commitment to expanding digital access and investing in comprehensive digital literacy training. This includes providing the infrastructure needed to connect underserved areas, such as affordable broadband, public Wi-Fi networks, and community tech hubs. It also means offering training programs that go beyond the basics, equipping residents with the skills needed to engage in higher education, access online certifications, and participate in digital economies.

Specific programs illustrate how these ideas translate into practice. For example, the National Digital Inclusion Alliance runs Digital Navigator programs across the U.S., helping marginalized residents access broadband, devices, and digital skills. In Colombia, "Amazonia 2.0" trains Indigenous leaders in remote sensing and participatory GIS to protect rainforests and improve resilience against climate impacts (Rodrigues et al., 2020). Such initiatives show how pairing technology with community leadership can accelerate resilience outcomes.

The Need for Community-Centric Digital Literacy Programs

To effectively address the digital divide and enhance resilience, digital literacy programs must be tailored to the specific needs and contexts of the communities they serve. Programs should not be one-size-fits-all; they should reflect the diverse realities of rural, urban, low-income, and minority communities. Imagine the opportunities created if digital literacy training in rural areas focused on skills that support agriculture, such as using mobile apps for weather forecasting or drone technology for crop monitoring. In urban areas, training might emphasize skills in tech entrepreneurship, digital marketing, or coding climate apps.

These initiatives should be designed with inclusivity at their core. This means actively recruiting participants from marginalized communities, providing training materials in multiple languages, and ensuring that instruction is accessible to individuals with disabilities. Training should also include discussions about digital ethics, privacy, and data security, empowering participants to navigate the digital world safely and confidently.

Programs that integrate climate literacy into digital skills training can further enhance resilience by helping participants understand how their new skills can be applied to address local environmental challenges. For instance, training that includes modules on how to use digital mapping tools, interpret climate data, or engage with digital advocacy platforms can empower communities to take proactive steps in resilience planning. By combining digital and climate education, these programs can build a foundation for long-term resilience and community empowerment.

The Path Forward: Linking Digital Access to Social Cohesion

As we continue to explore the role of digital access in resilience building, it is important to recognize that technology alone is not a panacea. The effectiveness of digital tools in supporting resilience depends on the social cohesion and community ties that underpin collective action. Digital access can enhance resilience by facilitating communication, coordination, and collaboration, but it is the strength of community relationships that ultimately determines how effectively these tools are used.

Climate change increasingly threatens communities worldwide, not only through direct impacts such as extreme weather events but also by exacerbating existing social and economic inequalities. Building resilience requires more than robust infrastructure; it necessitates a focus on social cohesion and economic mobility. The interplay between these factors creates a complex web that determines how communities can respond to and recover from climate-related shocks. By connecting the dots between digital access, technology, and social cohesion, we can build a holistic understanding of what it means to be resilient in a rapidly changing world.

The Role of Social Cohesion in Community Resilience

Social cohesion refers to the strength of relationships and the sense of solidarity among members of a community. It is a critical factor in community resilience, as cohesive communities are better equipped to respond to, recover from, and adapt to climate impacts. The National Academies of Sciences, Engineering, and Medicine (NASEM) emphasize that social cohesion significantly influences resilience, particularly in the context of hazard mitigation. Social networks and community bonds enhance the capacity for collective action, enabling communities to organize, share resources, and support one another during crises (NASEM Work on Social Cohesion and Hazard Mitigation).

During extreme weather events such as hurricanes, floods, and wildfires, cohesive communities demonstrate more effective responses. For example, neighbors often help each other evacuate, share critical information, and provide emotional support, which reduces the overall impact of disasters and facilitates faster recovery (NASEM). However, the success of social cohesion as a resilience strategy depends on inclusivity. Communities with strong social ties may still struggle if those ties are not extended to marginalized groups, including low-income families, immigrants, and people with disabilities. Ensuring equity in resilience planning is essential to avoid exacerbating existing social divides.

Digital inequities further complicate the role of social cohesion in resilience. Access to digital platforms for communication, coordination, and information dissemination is crucial during emergencies. Yet, the digital divide, which is marked by disparities in Internet access, digital literacy, and the availability of technological resources, limits the effectiveness of social cohesion strategies (NASEM Work on Social Cohesion and Hazard Mitigation). Communities without adequate digital infrastructure face challenges in organizing and responding effectively during disasters, often leaving the most vulnerable populations disconnected from vital information and support networks. Addressing the disconnect and the disadvantages associated with it requires resources, both external funding and assistance, as well as economic mobility in communities.

Economic Mobility as a Foundation for Resilience

Economic mobility, which is the ability of individuals and families to improve their economic status over time, is a critical foundation for resilience. When communities enjoy high levels of economic mobility, individuals are better able to invest in their futures, adapt to changing circumstances, and recover from setbacks. In contrast, low economic mobility traps people in cycles of poverty and vulnerability, significantly reducing their capacity to respond to and recover from climate impacts. According to the OECD's report *A Broken Social Elevator*, climate risks disproportionately affect those with limited economic mobility, making it harder for vulnerable populations to achieve upward mobility and build resilience (OECD, 2018). Climate resilience cannot be achieved without addressing the barriers that limit economic advancement.

The impacts of climate change exacerbate these existing inequalities, particularly in low-income communities. Extreme weather events such as floods, hurricanes, and wildfires often cause the most damage in areas with inadequate infrastructure and limited access to insurance. This lack of safety nets forces residents to deplete their savings or incur debt just to rebuild, as documented in both our study and previous literature (Wood et al., 2024). The result is a vicious cycle: the individuals and families most affected by climate disasters are also the least equipped to recover and adapt, thereby further limiting their economic mobility and resilience. The digital divide exacerbates these challenges by limiting access to online resources, financial services, and information that could aid in recovery. For example, communities without reliable

Internet access may miss critical early warnings or be unable to apply for disaster aid or job opportunities, further deepening their vulnerability.

In communities prone to flooding, the absence of strong infrastructure, such as levees or proper drainage systems, magnifies the destructive power of storms. Low-income residents often live in the most flood-prone areas because of the affordable but risky housing available to them. Without sufficient income, they are not only more exposed to hazards but also less able to recover. Our study identified income as the primary determinant of resilience, with a notable gap between communities with stable, salaried, or high-paying wage work and those where wages are stagnant or too low to cover basic needs such as rent, healthcare, or emergency savings (Wood et al., 2024). When rental costs exceed one-third of an individual's income, as we found in our research, eviction rates soar, and reliance on social safety nets increases. This precarious living situation is a direct barrier to economic mobility, and when disasters strike, it leaves already vulnerable populations with fewer resources to recover.

The Role of Income and Job Security

Our research on corporate community resilience models (CCRM) highlights the central role income plays in building resilience. Income was found to be 4.2 times more influential than access to healthcare in terms of predicting an individual's resilience (Wood et al., 2024). With higher income comes the ability to absorb financial shocks, rebuild homes, and access critical services such as healthcare. For example, wealthier individuals are better able to afford comprehensive insurance, which can be crucial for rapid post-disaster recovery. On the other hand, those earning low wages often lack access to insurance or face high out-of-pocket costs when attempting to recover from damages caused by disasters. This is especially true for renters, who often do not have the protections that homeowners do. According to our findings, there is no location in the U.S. where minimum-wage workers can afford to rent a two-bedroom apartment, highlighting how low income directly translates into heightened vulnerability during climate events.

The importance of income security cannot be overstated. Job stability and wage growth allow individuals to save for emergencies, invest in safer housing, and participate in community resilience efforts. As our study pointed out, higher income not only buys better access to physical protections (like home ownership or insurance) but also enables individuals to focus on long-term resilience rather than daily survival. Income stability is thus a critical buffer against the cascading effects of climate change. Our research suggests that policies to increase income—such as raising the minimum wage, providing cash assistance, or offering job training in higher-paying sectors—can significantly enhance resilience, especially in marginalized communities (Wood et al., 2024). It is important to acknowledge that economic mobility in today's world is closely tied to digital access. The digital divide hinders upward mobility by restricting opportunities for remote work, education, and access to emerging sectors such as the green economy, where digital skills are increasingly necessary.

The COVID-19 pandemic highlighted the interconnectedness of income security and resilience. As many workers were forced into unemployment or furloughs, those with savings were better able to endure the disruptions, while those living paycheck to paycheck faced immediate hardship. This is reflective of how climate disasters can disrupt economic livelihoods, disproportionately affecting those without a financial cushion.

Access to Healthcare and Job Benefits

Access to healthcare, which we found to be the second most important factor influencing resilience, is particularly critical for vulnerable populations. Lower-income individuals often face greater barriers to healthcare access, including the inability to afford routine or emergency care. Our study noted that nearly two-thirds of rent-burdened families in the U.S. have less than $400 in savings. In the event of a medical emergency, such as an injury sustained during a disaster, these families are often left to cope without the necessary resources for recovery. The digital divide compounds these challenges, as many healthcare services are moving online, making it even more difficult for individuals without reliable Internet access to receive timely care or access mental health resources that could aid in recovery from climate-related trauma.

Access to healthcare also has mental health implications. Disasters are traumatic, and the stress of losing one's home, income, or sense of security can exacerbate existing mental health conditions or create new ones. Communities that lack access to mental health services are thus less equipped to recover from the psychological impacts of climate-related events. Our study highlights the need for more comprehensive healthcare policies that include mental health support as a key component of resilience building (Wood et al., 2024). Providing affordable, accessible healthcare to low-income individuals can improve not only their overall well-being but also their ability to recover more quickly from disasters.

Job benefits, while often overlooked, also play a crucial role in building individual and community resilience. Beyond healthcare, job benefits such as paid leave, retirement plans, and professional development opportunities reduce the financial burden on workers during times of crisis. In our study, job benefits were ranked third in importance for resilience (Wood et al., 2024). Offering paid time off, for example, allows individuals to take time to rebuild their homes after a disaster without the fear of losing their job or income. This reduces stress and provides workers with the time and flexibility they need to recover effectively. Employers can also support resilience by providing benefits such as paid education and training programs that prepare workers for higher-paying, more resilient job sectors, especially in digitally connected fields that are increasingly important for future resilience.

As we discovered, many employers are beginning to recognize the importance of offering a more comprehensive benefits package, particularly in light of the pandemic. Paid parental leave, flexible work schedules, and other benefits are no longer seen as perks but as necessities that can enhance worker productivity and resilience. Companies that prioritize these benefits not only contribute to their employees' well-being but also improve overall community resilience by supporting economic mobility.

Policy Recommendations for Enhancing Economic Mobility and Resilience

Addressing economic mobility is essential for building climate resilience, particularly for vulnerable populations. Policymakers should require every federally funded climate resilience project to include a digital equity component, such as training programs, infrastructure upgrades, or local tech hubs. States could designate a percentage of adaptation funding specifically for advanced digital literacy, ensuring communities are prepared for emerging technologies like AI and blockchain. Additionally, partnerships between local governments, universities, and nonprofits can deliver tailored training that bridges both digital and climate knowledge gaps. These measures would ensure that technological advancements are integrated into resilience planning in ways that promote equity and community empowerment.

Our research suggests that targeted interventions like investments in education, job training, and affordable housing are critical to improving economic prospects for low-income communities while also enhancing their capacity to adapt to climate risks (Wood et al., 2024). These interventions must be inclusive and focus on dismantling systemic barriers that perpetuate inequality, including the digital divide that limits access to the very tools and opportunities needed for economic mobility. Expanding affordable broadband access and ensuring digital literacy are essential components of any strategy aimed at improving resilience and mobility, as the ability to engage with digital platforms is increasingly tied to economic success.

For instance, supporting education and job training in climate-resilient sectors, such as green energy, can provide individuals with pathways to higher-paying jobs that are less susceptible to the impacts of climate change. This aligns with our study's findings that job stability and income are the most critical factors in fostering resilience (Wood et al., 2024). Investing in affordable housing in less hazardous areas can also prevent vulnerable populations from living in high-risk zones, such as floodplains, thereby reducing their exposure to climate hazards.

Public and private sector collaboration is essential to implementing policies that foster economic mobility and resilience. Employers play a crucial role in offering job security and benefits that enhance resilience, while governments provide the regulatory framework and incentives to support these efforts. Public–private partnerships can drive innovation in areas such as job creation, healthcare access, digital inclusion, and housing affordability, which are critical for building community resilience. These partnerships not only leverage resources but also help to create holistic solutions that address the root causes of vulnerability. Ensuring equitable access to these opportunities, particularly through digital inclusion, will be key to creating resilient, adaptable communities capable of withstanding future climate shocks.

Economic mobility and social cohesion are deeply interconnected and mutually reinforcing. Communities that foster economic mobility often benefit from stronger social bonds, as individuals have greater opportunities to participate in economic and social life. Conversely, social cohesion can bolster economic mobility by creating networks of support, mentorship, and access to resources that help individuals climb the economic ladder (OECD, 2018). In times of crisis, these connections make the difference between recovery and prolonged hardship. In communities where social cohesion is strong, residents frequently pool resources, share information, and support each other through challenges. These social networks act as informal safety nets, providing critical support during times of crisis, and their contributions to resilience are invaluable.

When economic mobility is limited, social cohesion can weaken. Economic hardship strains social bonds, leading to increased stress, conflict, and a sense of disconnection. This is particularly evident in communities that experience repeated climate-related shocks without sufficient resources to recover. Addressing the underlying economic challenges is essential for maintaining and strengthening social cohesion. Social cohesion, in turn, supports resilience by enabling communities to respond more effectively to disasters and reduce the need for external assistance (NASEM Work on Social Cohesion and Hazard Mitigation). Bridging the digital divide can help strengthen these communal ties, as access to digital tools and platforms fosters greater communication, resource sharing, and community-led resilience efforts.

Digital Divide and the Challenges of Inclusive Resilience

The digital divide poses a significant barrier to both social cohesion and economic mobility, directly impacting community resilience. As digital technologies become increasingly integral to climate adaptation and disaster response, the lack of access to these technologies among marginalized communities undermines their resilience. The integration of advanced data technologies, such as big data, machine learning, and sensor networks, into nature-based solutions has the potential to revolutionize climate adaptation by providing real-time data for decision-making and resource management (Harnessing New Data Technologies for Nature-Based Solutions). However, these benefits are not evenly distributed.

Disparities in access to data and analytical tools between developed and developing regions limit the ability of vulnerable communities to participate in and benefit from data-driven resilience strategies. Rural areas, low-income neighborhoods, and communities of color often face significant challenges in accessing digital technologies, including high costs, limited Internet connectivity, and inadequate digital literacy. These digital inequities reinforce existing vulnerabilities, leaving marginalized populations without the resources they need to prepare for and respond to climate impacts.

The consequences of the digital divide extend beyond immediate disaster response. In the long term, limited access to digital technologies hampers economic mobility by restricting access to education, job opportunities, and financial services (Harnessing New Data Technologies for Nature-Based Solutions). For example, remote learning platforms and online job applications are less accessible to those without reliable Internet access, putting residents at a disadvantage in an increasingly digital economy. Bridging the digital divide is therefore critical for enhancing both resilience and mobility, ensuring that all members of society have the opportunity to thrive in a changing climate.

Digital technologies have the potential to transform community engagement and disaster preparedness when they are accessible. Tools such as mobile apps, social media, and geographic information systems

(GIS) enable communities to monitor environmental changes, communicate risks, and coordinate response efforts in real time. The uneven distribution of these technologies often results in disparities in resilience outcomes, leaving some communities better equipped than others to respond to and recover from disasters. Investments in digital infrastructure, combined with targeted training and outreach programs, are essential to bridging this gap. The insights from surveys of emergency managers and climate resilience officers, explored in the next chapter, underscore the critical role of digital access in equitable resilience-building and highlight the practical challenges and opportunities in implementing these solutions effectively. (Harnessing New Data Technologies for Nature-Based Solutions).

Community Engagement and Inclusive Decision-Making

Building resilience through social cohesion and economic mobility requires actionable steps to engage communities and ensure inclusive decision-making processes. Practical strategies include fostering a sense of ownership among residents, empowering them to participate in resilience planning, and tailoring interventions to meet the unique needs of marginalized groups. This collaborative approach helps communities not only prepare for and respond to disasters but also strengthen their long-term adaptability and sustainability.

To effectively engage communities, participatory approaches should prioritize co-creation of solutions, bringing together residents, local governments, and other stakeholders. For example, hosting regular workshops and public forums can ensure that community members have a platform to share their knowledge and contribute to decision-making. Recognizing and integrating local knowledge alongside modern technologies can lead to innovative and effective resilience strategies. These strategies, as demonstrated in initiatives such as the government-to-sovereign Indigenous nation climate relocation case discussed later, have been successful in protecting ecosystems, reducing disaster risks, and enhancing digital access and skills for broader resilience building (NASEM Work on Social Cohesion and Hazard Mitigation).

Overcoming systemic barriers to inclusive decision-making is critical for equitable resilience. Practical measures include providing translation and accessibility services for non-English speakers, addressing logistical barriers like transportation to community meetings, and intentionally including representatives from marginalized groups in decision-making bodies. Digital tools can further enhance these efforts by making resilience planning more accessible. For instance, interactive online platforms can gather community input, track environmental changes, and disseminate critical information.

Case studies in Chapter 6 will illustrate how communities have successfully used digital access, skills, and collaboration to drive resilience. These examples underscore the importance of inclusive, technology-supported engagement in achieving sustainable outcomes. When communities are given the tools and spaces to lead resilience efforts, they not only overcome immediate challenges but also build trust, strengthen social cohesion, and create economic opportunities, all of which contribute to their long-term resilience. (NASEM Work on Social Cohesion and Hazard Mitigation).

True Resilience Building: Possibilities Through Combined Solutions

True resilience building requires combining the strategies of digital access, nature-based solutions (NBS), and inclusive decision-making to address existing inequalities and prevent future marginalization. By resolving systemic barriers and leveraging innovative technologies, communities can build resilience that not only addresses current vulnerabilities but also anticipates and mitigates future risks. This holistic approach ensures communities are equipped to navigate environmental, social, and economic challenges in an increasingly complex world.

The intersection of technology and NBS represents a powerful framework for resilience. Technological innovations such as remote sensing, geospatial analysis, and renewable energy systems enhance our ability to understand and manage environmental challenges like climate change and natural disasters. Smart infrastructure, for instance, integrates sensors and automation to optimize the efficiency of NBS, such as flood-resilient wetlands or heat-mitigating urban forests. These technologies, when combined with traditional knowledge and local expertise, create robust and adaptable solutions that align with the needs of diverse communities.

Technologies like remote sensing, Geographic Information Systems (GIS), and data modeling have become invaluable tools in the design and implementation of NBS. Remote sensing allows for the monitoring of large-scale environmental changes, such as deforestation, coastal erosion, and flooding, providing critical data that informs adaptive management strategies. GIS can map vulnerable areas, identify priority zones for intervention, and simulate the effects of different management scenarios, helping communities make informed decisions about where and how to implement NBS.

The integration of advanced data technologies, including big data analytics, machine learning, and sensor networks, further enhances the effectiveness of NBS by providing real-time data for adaptive management. Sensor networks can monitor water levels in rivers and reservoirs, providing early warnings of flood risks. Machine learning algorithms can predict climate impacts on ecosystems, enabling proactive measures to protect vulnerable habitats.

These technologies also support the restoration of degraded ecosystems by guiding reforestation efforts, optimizing the placement of green infrastructure, and assessing the success of restoration projects over time. For instance, machine learning models can analyze historical weather data and project future climate scenarios, guiding the planting of tree species that are more resilient to changing conditions.

Despite their potential, the deployment of technology-supported NBS faces significant challenges, particularly related to the digital divide. Access to digital tools, data, and technical expertise remains uneven, especially in rural and underserved areas. Limited access to high-quality data and advanced analytics hampers the ability of these communities to fully benefit from technology-supported NBS. Additionally, the cost of implementing these technologies can be prohibitive, particularly for low-resource settings, making it difficult to scale successful interventions.

To maximize the benefits of emerging technologies like blockchain, dApps, and DeFi, communities need robust digital literacy and access programs. These technologies hold promise for enhancing community resilience by offering new ways to manage resources, facilitate transactions, and fund community-led initiatives. Without targeted education on how to navigate and utilize these tools, the potential remains out of reach for many. Digital literacy programs should include training that demystifies Web 3.0 platforms and provide practical skills to engage with decentralized systems effectively.

This knowledge empowers communities not only to use these tools but also to participate actively in shaping the digital landscape to meet local resilience needs. Bridging the digital divide with comprehensive training in advanced technologies also creates pathways to economic mobility, addressing both the systemic barriers that exclude marginalized groups and the evolving needs of the digital economy. This integration of technology with community-led resilience strategies ensures that digital transformation supports social and economic empowerment. It is important for policymakers to understand these intersections and design programs and mechanisms that align with the experiences of these communities.

Web 3.0, AI Tools, and Decentralized Approaches to Resilience

Web 3.0 represents the next stage of Internet evolution, marked by decentralization, distributed data ownership, and direct peer-to-peer engagement. This shift is opening new ways for communities to participate in resilience building, climate action, and disaster risk reduction. Emerging technologies such as blockchain and artificial intelligence (AI) offer opportunities to democratize data access, create innovative funding models, and increase public participation in climate-related decisions.

Current debates in climate governance increasingly explore the role of advanced digital technologies in adaptation and resilience. AI's ability to analyze large datasets and generate predictive models holds potential to improve how communities understand and respond to climate risks. As LaMadrid and Clair (2017) explain, digital systems can either expand democratic participation in environmental governance or reinforce existing divides if access remains uneven.

Web 3.0 also creates space to embed participatory governance within digital infrastructures. Technologies such as blockchain and decentralized applications (dApps) can help record climate actions, facilitate community-led funding, and increase transparency in resilience initiatives. However, without inclusive design and capacity-building, these systems may benefit only well-resourced users. Many communities still lack the infrastructure or training to fully participate.

AI governance frameworks like the OECD AI Principles and UNESCO's Recommendation on the Ethics of AI emphasize transparency, accountability, and fairness. These principles are essential to ensure that resilience technologies protect community data and promote trust in climate tools (OECD, 2019; UNESCO, 2021). Integrating such ethical standards into adaptation planning helps advance equitable use of technology and ensures that digital systems reflect local values and needs.

Blockchain technology supports many Web 3.0 innovations. It offers a secure and transparent way to manage data, track environmental impacts, and distribute funds. In climate resilience work, blockchain systems can:

- Track carbon credits and verify sustainability goals.
- Monitor environmental indicators in real time.
- Ensure transparent allocation of climate adaptation funds.
- Support ethical supply chains for nature-based solutions (Schindler & Wolf, 2020).

Blockchain's decentralized ledgers reduce the need for intermediaries, cutting costs and limiting opportunities for corruption. Smart contracts, which automatically execute agreements based on pre-set conditions, can be used to release funding when climate adaptation milestones are met. These mechanisms reduce administrative burden and improve accountability.

Decentralized applications (dApps) extend blockchain's functions by creating interactive tools for civic participation. Communities can use dApps to report local climate impacts, such as flooding or extreme heat, which builds real-time environmental databases. In some cases, residents may vote on resilience priorities using blockchain-based platforms. This reflects the idea of "polisdigitocracy," where digital platforms allow residents to shape environmental decisions (LaMadrid & Clair, 2017). These tools support the co-creation of solutions and amplify community knowledge. However, their successful use depends on digital literacy and trust in the system. In areas impacted by historical disinvestment or information redlining (Dorries & Rhoads, 2021), additional outreach and training are essential to support adoption.

Decentralized finance (DeFi) uses blockchain to provide peer-to-peer financial services without traditional banking systems. In resilience contexts, DeFi tools can:

- Support local crowdfunding for projects like microgrids or reforestation.
- Provide microloans for home retrofits or flood prevention.
- Offer incentive-based systems for environmental stewardship.

DeFi can expand access to funding in communities that have been historically excluded. Still, new financial tools carry risks. Without strong governance, these systems may become targets for fraud or replicate wealth disparities in digital spaces.

While blockchain promotes transparency, AI and web-based platforms enhance communication and awareness. These systems can translate climate science into community-relevant tools, such as:

- Localized risk maps and emergency alerts.
- Dashboards to visualize climate projections or flood zones.
- Feedback channels where residents report impacts or suggest responses.

AI can also analyze real-time data from social media or sensors, detecting disasters before official systems react. However, Linnenluecke et al. (2017) caution that digital tools can fail without trust or cultural relevance. Effective climate communication must move beyond one-way alerts and include dialogue with communities.

Toward an Equitable Digital Resilience Future

To enhance resilience through social cohesion, economic mobility, and technology integration, a multifaceted approach is necessary. First, investments in social and digital infrastructure are crucial to bridging gaps in access to critical resources. By funding community centers, social services, and local nonprofits, governments and private sector actors can strengthen the social networks that are foundational to resilience. At the same time, providing broadband access and investing in advanced technologies like satellite networks and sensor systems will ensure that rural and underserved communities have the tools needed to stay connected and prepared, particularly during disasters.

These efforts should be complemented by public–private partnerships that drive innovation in technology, allowing for the development of AI-powered monitoring systems and blockchain-based environmental management tools that improve resilience at scale. Equipping communities with technical capacity and digital literacy is equally essential. Targeted training programs must prioritize skills in data collection, analysis, and decision-making, empowering communities to leverage technologies like Web 3.0 in their resilience efforts. By ensuring that these skills reach traditionally excluded populations, such as low-income families and communities of color, governments and the private sector can foster economic mobility and adaptive capacity.

Education and job training initiatives focused on green jobs and climate-resilient sectors further support these goals by providing pathways to higher-paying roles, enabling vulnerable populations to participate in the growing digital economy while building resilience to climate impacts. Implementing these strategies will require close collaboration with community leaders and leveraging existing federal and state programs, such as adaptation and hazard mitigation funds. Most of the digital infrastructure and skills development resources lie within the private sector, underscoring the importance of fostering public–private–community partnerships for effective problem-solving.

These collaborative efforts will help ensure that technological advancements are accessible to all and integrated into local resilience strategies in ways that promote long-term sustainability and equity. We explored the complexities of building holistic resilience in the digital era, emphasizing the intersections between digital access, social cohesion, economic mobility, and technology-supported solutions. Climate challenges demand an integrated approach that considers the social, economic, and technological factors shaping community resilience.

The digital divide remains a significant barrier, disproportionately affecting marginalized communities and limiting their participation in resilience-building efforts. Equitable digital access goes beyond Internet connectivity. It provides the tools, skills, and opportunities individuals need to leverage digital resources for education, employment, and resilience planning. Comprehensive digital literacy programs must address both basic and advanced skills, offering pathways into the digital economy and equipping participants to interpret climate data, engage with technological solutions, and advocate for their needs.

Addressing the digital divide requires tackling historical inequities in digital upskilling programs to ensure that all communities have access to meaningful digital education that leads to technology-driven careers. Integrating climate literacy into digital skills training empowers communities to understand and manage their unique climate risks, transforming digital tools into catalysts for resilience.

Social cohesion remains a critical focus as we move forward. Strong community ties and collective capacity to respond to crises are foundational to resilience and must be nurtured alongside technological advancements. Digital access can enhance these connections, facilitating communication and coordination during emergencies, but it is ultimately the human relationships and shared values that drive community action and recovery.

Building holistic resilience requires intentional actions, inclusive policies, and a commitment to addressing inequality. It involves reimagining strategies that prioritize equity, access, and empowerment while recognizing the interconnectedness of social, economic, and technological factors. The next chapter, "Bridging the Digital Divide for Climate Resilience," will delve deeper into practical steps for achieving digital justice and climate resilience, drawing insights from those working in these communities and offering practical steps that can be taken to ensure digital justice and climate resilience.

References

Ajiboye, J. O., & Adebayo, F. A. (2022). Digital learning: Key to bridging the skills gap in the Nigerian experience. In *Emerging trends in educational technology* (pp. 85–102). Springer. https://link.springer.com/chapter/10.1007/978-981-16-3521-2_6

Aspen Institute. (2024). *Roads to resilience: Rural America and climate change*. The Aspen Institute. https://www.aspeninstitute.org/events/roadsclimate/

Atallah, D. G., Bacigalupe, G., & Repetto, P. (2021). Centering at the margins: Critical community resilience praxis. *Journal of Humanistic Psychology, 60*(3), 295–315. https://journals.sagepub.com/doi/abs/10.1177/0022167818825305

Attygalle, K., & Jayasuriya, S. (2021). *Climate change and community resilience: Insights from South Asia*. Springer. file:///Users/ms4532/Downloads/9789811606809.pdf

Bertilsson, L., Wiklund, K., de Moura Tebaldi, I., Rezende, O. M., Veról, A. P., & Miguez, M. G. (2018). Urban flood resilience—a multi-criteria index to integrate flood resilience into urban planning. *Journal of Hydrology, 573*, 970–982. https://doi.org/10.1016/j.jhydrol.2018.06.052

Cai, H., Lam, N. S., Qiang, Y., Zou, L., Correll, R. M., & Mihunov, V. (2018). A synthesis of disaster resilience measurement methods and indices. *International Journal of Disaster Risk Reduction, 31*, 844–855. https://doi.org/10.1016/j.ijdrr.2018.07.015

Chuang, W. C., Garmestani, A., Eason, T. N., Spanbauer, T. L., Fried-Petersen, H. B., & Roberts, C. P. (2018). Enhancing quantitative approaches for assessing community resilience. *Journal of Environmental Management, 213*, 353–362. https://doi.org/10.1016/j.jenvman.2018.01.083

Cretney, R., & Bond, S. (2019). *Centering at the margins: Critical community resilience praxis*.

Cunsolo, A., & Ellis, N. R. (2021). Technology in support of nature-based solutions. *Ecology and Society, 26*(4), 35. https://www.ecologyandsociety.org/vol26/iss4/art35/

Dahl, J., & Bell, D. (2021). *Building resilient communities over time*. TU Delft Repositories. https://research.tudelft.nl/en/publications/building-resilient-communities-over-time

Dorries, H., & Rhoads, R. (2021). Information redlining: Mapping disparities in digital information access. *Journal of Library Administration,*

61(3), 197–210. https://www.tandfonline.com/doi/abs/10.1080/01930826.2021.1906559

Heijnen, S. (2020). *Leveraging digital development for community resilience*. Queensland University of Technology ePrints. https://eprints.qut.edu.au/211565/

Horney, J., Dwyer, C., Chirra, B., McCarthy, K., Shafer, J., & Smith, G. (2018). Measuring successful disaster recovery. *International Journal of Mass Emergencies and Disasters, 36*(1), 1–22. https://doi.org/10.1111/disa.12190

Houston, J. B., Spialek, M. L., Cox, J., Greenwood, M., & First, J. (2015). The effect of social media on the dynamics of (self) resilience during disasters: A literature review. *Disasters, 39*(1), 1–22. https://onlinelibrary.wiley.com/doi/full/10.1111/1468-5973.12212

LaMadrid, A., & Clair, A. (2017). Polisdigitocracy: Citizen engagement for climate action through digital technologies. *ResearchGate*. https://www.researchgate.net/publication/318092081_Polisdigitocracy_Citizen_Engagement_for_Climate_Action_through_Digital_Technologies

LINK Health. (2024, July 16). *Bridging the digital divide in rural America*. https://link-health.org/2024/07/16/bridging-the-digital-divide-in-rural-america/

Linnenluecke, M. K., Griffiths, A., Winn, M. I., & McKnight, B. (2017). Communicating climate change adaptation information using web-based platforms. *ResearchGate*. https://www.researchgate.net/publication/318431221_Communicating_climate_change_adaptation_information_using_web-based_platforms

Metz, R. (2023, February 15). Here's what the sustainable cities of tomorrow could look like. *Wired*. https://www.wired.com/story/heres-what-the-sustainable-cities-of-tomorrow-could-look-like/

Moghadas, M., Asadzadeh, A., Vafeidis, A., Fekete, A., & Kötter, T. (2019). A multi-criteria approach for assessing urban flood resilience in Tehran, Iran. *International Journal of Disaster Risk Reduction, 35*, Article 101069. https://doi.org/10.1016/j.ijdrr.2019.101069

Nelson, D. R., Bledsoe, B. P., Ferreira, S., & Nibbelink, N. P. (2020). Challenges to realizing the potential of nature-based solutions. *Current Opinion in Environmental Sustainability, 45*, 49–55. https://doi.org/10.1016/j.cosust.2020.09.001

OECD. (2018). *A broken social elevator? How to promote social mobility*. OECD Publishing. https://www.oecd.org/social/soc/Social-mobility-2018-Overview-MainFindings.pdf

OECD. (2019). *OECD principles on artificial intelligence*. OECD Publishing. https://www.oecd.org/going-digital/ai/principles/

Okolloh, O. (2009). Ushahidi, or 'testimony': Web 2.0 tools for crowdsourcing crisis information. *Participatory Learning and Action, 59*(1), 65–70. https://pubs.iied.org/14607iied

Rahman, M. H., & Mutum, D. S. (2021). *A ethnographic study of the social media influencers on creating netizens' climate change awareness on Instagram*. SSRN. https://www.academia.edu/download/83322032/SSRN_A_Netnography_Study_of_the_Social_Media_Influencers_on_Creating.pdf

Regnier, A. (2023, March 3). How cities are using nature-based solutions to tackle floods. *Time*. https://time.com/7202917/cities-nature-based-solutions-floods/

Rodrigues, A. S. L., Silva, R. A., Sánchez-Triana, E., & Varela, E. (2020). Integrating Indigenous knowledge into environmental monitoring and resilience planning in the Amazon. *Environmental Science & Policy, 112*, 48–58. https://doi.org/10.1016/j.envsci.2020.05.007

Sanders, M. (2023a). *Enhancing federal climate initiatives: Integrating tech-focused green jobs for equity and innovation*. https://fas.org/publication/enhancing-climate-initiatives-green-jobs/

Sanders, M. (2023b). *Using a digital justice framework to improve disaster preparation and response*. Federation of American Scientists. https://fas.org/publication/using-a-digital-justice-framework-to-improve-disaster-preparation-and-response/

Sanders, M. C., & Haughton, R. (2024). Digital divide and disasters: Challenges and opportunities in the age of climate change. *Journal of Disaster Studies*.

Schindler, M., & Wolf, T. (2020). Harnessing new data technologies for nature-based solutions: A review. *Environmental Science & Policy, 115*, 182–195. https://www.sciencedirect.com/science/article/pii/S2212420920312978

Serfilippi, E., & Ramnath, G. (2018). Resilience measurement and conceptual frameworks: A review of the literature. *Annals of Public and Cooperative Economics, 89*(4), 645–664. https://doi.org/10.1111/apce.12202

The National Academies of Sciences, Engineering, and Medicine. (2020). *Applied research topics for hazard mitigation and resilience*. The National Academies Press. https://www.nationalacademies.org/our-work/applied-research-topics-for-hazard-mitigation-and-resilience

UNESCO. (2021). Recommendation on the ethics of artificial intelligence. UNESCO. https://unesdoc.unesco.org/ark:/48223/pf0000381137

Wood, E., Lam, J. C., & Sanders, M. (2024). Economic mobility and resilience: Addressing inequities in a changing climate. *Journal of Emergency Management, 22*(7), 29–38.

World Bank. (2020). *Digital gender divide: Understanding the gender gap in internet access.* World Bank Group. https://openknowledge.worldbank.org/handle/10986/33952

3

Bridging the Digital Divide for Climate Resilience

If the social contract of earlier eras relied on shared investments in traditional infrastructure such as waterworks and bridges, today's interpretation is about broadband networks, data systems, economic opportunity, and emergency resources. Where the digital divide intersects with climate resilience efforts, it reflects a structural failure to extend the benefits of the modern social contract to all communities. It lacks a justice framing.

When we took a deeper look at how climate and digital justice interface, it made sense to begin in an area I know well, the Gulf Coast of the United States. This chapter pulls from research in our "Ten States" project to better understand the value of a digital justice framing of resilience and of democratizing climate change information.

Why the Digital Divide Flows with Disaster and Climate Risk

Why does the digital divide so often align with disaster and climate risk? There is a connection between the co-location of public goods in the built environment and bias in policy construction. In today's interconnected world, the digital divide remains a significant barrier, exacerbating

M. C. Sanders, *Climate and Code*, https://doi.org/10.1007/978-3-032-28032-9_3

existing inequalities across various domains. This divide is closely intertwined with both disaster and climate risk, forming a complex web of challenges that disproportionately affect marginalized populations (Fast Company, 2021a, 2021b). At the intersection of the digital divide and disaster/climate risk, a recurring pattern emerges: those with the least access to the Internet and the lowest digital literacy are often found in rural areas or among formerly redlined and marginalized communities in urban areas (Rothstein, 2017).

Using survey insights and interviews, I take a close and intentional look at the intersection of race and climate risk in the Gulf Coast of the United States. This region, with its history of racial disparities and vulnerability to extreme weather events, serves as a poignant example of how marginalized communities bear a disproportionate burden of climate impacts (Bullard, 2000; Rothstein, 2017). By understanding the interconnectedness of race, climate risk, and the digital divide, we gain valuable insights into the complex dynamics that perpetuate inequality and hinder resilience efforts.

Climate Change, the Digital Divide, and Societal Implications: Literature Review

Climate change, the digital divide, infectious diseases, and unemployment represent some of the most significant global risks facing society today, intertwined in complex and impactful ways. The digital divide complicates these interactions further and makes them global threats to social stability and economic security (Sanders, 2024a, 2024b, Fast Company, 2021a, 2021b). This frames the deeper investigation into how these risks are connected, particularly how the digital divide can amplify climate change's adverse effects. The digital divide refers not only to disparities in access to digital technologies but also to variations in capabilities to use these technologies effectively across populations.

Warschauer (2003), in *Technology and Social Inclusion*, argues that merely providing technology access is insufficient for true social inclusion. The focus should be on usability and integration into daily life, which enhances communities' abilities to engage with and respond to climate challenges. This perspective is vital for understanding why some communities show greater resilience when facing such adversities.

The study of Hurricane Maria's impact on Puerto Rico offers a clear example of how technology, specifically social media and digital communication, plays a vital role in managing climate-induced disasters (PNAS, 2018). The research analyzes population displacement resulting from Hurricane Maria using social media and digital phone data, compared with air travel and census records, to approximate changes in population size. The authors explain that population shifts translate into rural-to-urban migration due to damaged infrastructure and a lack of resources to rebuild communities. This study illustrates that while technology can track and analyze population movements and resilience, a lack of access to such technology hinders effective response and recovery, highlighting digital access's critical role post-disaster.

In *Digital Economies at Global Margins*, Unwin (2019) discusses the economic aspects of the digital divide, noting that marginalized or economically disadvantaged groups often have the least access to technology. Access to and understanding of the digital world improve both individual financial situations and broader economies. Unwin highlights that marginalized populations are not solely defined geographically; both urban and rural communities face digital exclusion alongside climate change challenges. This bridges the gap between climate change and digital access: climate change further separates many communities from digital inclusion, worsening their quality of life.

We talk about quality of life, hand in hand with sustainability. This is commonly framed through three interrelated pillars: economic, environmental, and social sustainability. Together, these pillars provide a useful framework for understanding how climate change affects the stability and well-being of societies (CASS, 2014). The authors emphasize that environmental conditions are closely tied to social sustainability, meaning the ability of communities to maintain quality of life and meet human needs over time. Their analysis shows that disruptions such as declining water quality, food scarcity, or pollution can undermine social stability and community cohesion. In particular, the study notes that "food, water, and energy will, directly and indirectly, affect important aspects of societal stability." Within this framework, social sustainability refers to the satisfaction of a broad set of human needs and the maintenance of conditions that allow people and communities to live well and thrive. This perspective highlights that the impacts of climate change extend beyond environmental damage to include consequences for the social fabric of communities. The framework of the three pillars is also relevant to understanding the digital divide. As environmental pressures intensify, inequalities in digital access and technological capacity may widen, further limiting the ability of vulnerable populations to respond to climate and disaster risks.

Addressing these issues requires multifaceted approaches that consider not only technology provision but also enhancement of capabilities. Natural disasters' effects on communication networks underscore infrastructure vulnerability to climate events (Yamori & Sugawara, 2017). Such disruptions can significantly impede disaster response and recovery, particularly in areas already struggling with digital inequity. Together, this literature provides a comprehensive view of how the digital divide and climate change intersect to affect social stability and community resilience.

The Intersection of Digital Divestment and Climate Risk: An Analysis of Co-location, Bias, and Infrastructure

As researchers recognize the connection between digital divestment and climate risk, the next step should be to understand co-location and bias. Co-location refers to geographic overlap between limited digital access and vulnerability to environmental hazards. This overlap is not coincidental but is deeply rooted in historical and structural biases that continue to affect contemporary resource allocation (Bullard, 2000; Rothstein, 2017). Studies, such as those by the journal "Climate" and research highlighted on platforms like Bloomberg and Human I-T, illustrate how rural areas and marginalized urban communities frequently suffer from insufficient investment in critical infrastructures, including broadband and climate adaptation mechanisms (Cooper, 2023).

Digital divestment, the lack or withdrawal of digital infrastructure investment, disproportionately affects communities that are also vulnerable to climate impacts. This divestment is often a result of historical redlining and systemic neglect, such as that implicated in the Indigenous reservation system, which has led to a concentration of vulnerabilities in these communities (Rothstein, 2017).

A key factor contributing to the correlation of digital divestment and climate change is the design of infrastructure. Divested communities are typically located away from infrastructures such as highways and urban centers, which are historically more likely to receive continuous investments in both digital and physical infrastructure. The design of national infrastructure, notably the interstate highway system, provides a clear example of how initial planning decisions have long-term impacts on accessibility to public goods. As noted in the literature, highways facilitated significant economic and social connectivity across the country, but also excluded certain communities (Bullard & Wright, 2012; Foster, 1998). This exclusion extends beyond physical connectivity; it influences where utilities and services, including Internet infrastructure, are located. Government agencies, utilities, and digital connectivity follow the paths carved by this highway system. Residents in distant areas, often rural or historically marginalized urban neighborhoods, face compounded disadvantages. These areas are less likely to have reliable Internet access, which is crucial not only for communication but also for accessing climate risk information and emergency services during disasters.

The relationship between digital access and climate resilience is also influenced by the availability of and capacity to interpret relevant data. Research published in *PLOS Climate* discusses machine learning's role in predicting Scope 3 emissions, highlighting advanced analytics' potential in understanding and mitigating climate risks. A downfall of these technologies is that they assume access to necessary digital tools and literacy to interpret outputs (Nguyen et al., 2023), despite this literacy being an essential component. In areas with limited educational resources and upskilling opportunities, populations are less equipped to utilize digital tools effectively. This limitation reduces their ability to engage with technologies and participate in digital economies, further exacerbating economic disparities.

Understanding the Digital Divide in Climate Vulnerable States

My team and I began a multi-year study in ten of the most climate-vulnerable U.S. states, which are also among the most digitally divested. These states, excluding New York and California, are home to about 60% of the African-American population. This combination highlights the gap between those with access to modern information and communication technology and those without. We examined how climate vulnerability, digital divestment, race, and political instability collide in Texas, Alabama, Mississippi, Louisiana, Florida, Georgia, South Carolina, North Carolina, Maryland, and Virginia (U.S. Census, 2021). These divisions not only affect general connectivity but also critically undermine community resilience to climate impacts.

Comparing the Gulf South and Midwest with more digitally inclusive regions underscores disparities in climate risk preparedness. The digital divide becomes especially visible when examining the role of social media and digital communication in disaster response and recovery. After Hurricane Maria, digital trace data helped researchers approximate displacement and population changes (Acosta et al., 2020). In practice, the ability to use these tools can alter a community's resilience trajectory.

These states, marked by susceptibility to hurricanes, floods, and severe weather, face unique challenges exacerbated by the digital divide. Areas with limited Internet access struggle to receive timely weather alerts, evacuation notices, and disaster relief information. The divide is not purely technological but also influenced by socioeconomic factors such as income disparities, educational levels, and urban–rural divides, which collectively contribute to unequal information and resource access.

Emergency response systems increasingly depend on digital communication. Approximately 80% of emergency calls come from wireless devices—a sign of dependency on cell phones (National Emergency Number Association, n.d.). Emergency alerts come over TV signals, streaming devices, radio, Wi-Fi, and well-established cell and broadband networks. This means that for those who do not have Internet at home, use pay-for plans, don't have smartphones or consistent access to them, the broadband gaps limit vulnerable populations' timely access to emergency services (Bullard, 2000; Sanders, 2024b). This impedes response and exacerbates the impacts of disasters. Bridging the digital divide is critical for equitable emergency access.

The digital divide affects not only emergency communication but also long-term recovery, climate adaptation, and mitigation. Post-disaster recovery relies heavily on online platforms for aid applications and information on rebuilding and funding. Those without adequate digital access face disadvantages in navigating aid processes, often experiencing delays in support. They are frequently excluded from virtual town halls, online forums, and digital decision-making on community planning against climate threats. These groups often lack representation in planning processes, resulting in adaptation strategies that do not fully address their vulnerabilities or needs.

Survey Insights and Analysis

In these states, the digital divide intersects with racial and ethnic disparities. Historically marginalized communities, including many Black and Hispanic populations in the South, often experience both higher climate vulnerability and lower Internet access. This intersection compounds challenges during climate events, where a lack of timely, relevant information exacerbates inequalities and hinders response and recovery. Recognizing the intersectionality of race and climate risk allows targeted interventions addressing digital divides and systemic inequities exacerbating climate vulnerabilities. To deepen our understanding of these impacts across the region, we surveyed emergency managers, planners, and chief resilience officers in these ten states.

Survey Methodology

The survey targeted a specific group of professionals: emergency managers, planners, and chief resilience officers across ten of the most climate-vulnerable states in the U.S. The selection of these states was guided by their historical and projected climate-related challenges. The survey aimed to gauge the awareness and understanding of these professionals regarding the intersection of the digital divide and climate vulnerability. A total of 120 individuals holding titles such as "emergency manager," "resilience officer," and "disaster specialist" were contacted, with a structured questionnaire that posed initial and follow-up questions based on their responses. The pivotal question asked was, "Are you aware that communities impacted by the digital divide are also more vulnerable to the impacts of climate change?" Those who responded affirmatively were then asked to elaborate on their understanding of this connection.

Key Findings

Emergency managers noted the digital divide complicates efforts to communicate effectively with all residents, especially those in rural or economically disadvantaged areas who may not regularly use the Internet. Many highlighted that a significant portion of the population lacks reliable Internet, crucial for timely weather updates and emergency service access. This digital connectivity gap hampers preparedness and response, complicating vital information dissemination during critical times. The survey revealed a nuanced understanding among emergency managers of how the digital divide exacerbates climate vulnerability. A significant majority acknowledged the divide as a critical factor influencing community resilience. Insights underscored the consensus that communities lacking robust digital infrastructure are disadvantaged in accessing timely weather threat and disaster recovery information.

Quotes and Real-Life Examples

The survey responses were punctuated with compelling testimonials that illuminated the challenges and perspectives of the respondents. One emergency manager from a state known for its "wide" digital divide noted:

> Yes. Such communities do not have access to existing and future data concerning historical and climate-related risks... Our state has a wide digital divide, and one of our pressing issues is communicating with folks who don't have or use the internet regularly.

Another highlighted broader implications:

> Yes, communities without internet access are less likely to access critical information and can get 'lost' when agencies try outreach if they are not intentional about reaching these communities. The digital divide correlates with a lack of access to other services, compounding vulnerabilities.

Keeping with the research pool of the same titles, we probed further, asking, "Are you aware that communities impacted by the digital divide are also more vulnerable to the impacts of climate change?" If their answer was "yes," we asked the respondents to explain their understanding of the connection.

> One respondent noted: "Yes. The digital divide magnifies vulnerability to all hazards. For climate change in particular, it includes less or no access to warnings, prep/safety info, access to resources, opportunity for input or to request assistance, and likely correlation with less adaptable housing and infrastructure."

When asked specifically about the relationship between the digital divide, climate change, and rural and BIPOC communities, the response was split, with many in these roles not understanding or being aware of the connection. For those that did, the responses ranged from having an emergency manager become part of the local digital equity committee to one person using hazard mitigation funding to complete a local broadband project.

Only two respondents had the opportunity to collaborate with state or local officials to understand, resource, or otherwise address the intersection of the digital divide and climate change. However, when asked about the "obstacles you encounter when addressing the digital divide and climate risk," they weighed in heavily. Many of the responses referred to a lack of political will or political leadership. There were also concerns about the resource process for both issues, and certainly the confluence of the two being politicized. Others cited a need to have hazard mitigation funding consistently available for this issue. Outside of the larger broadband access initiatives, they believe that digital communications infrastructure, local Wi-Fi, or other kinds of connectivity projects should be categorized as "allowable uses" of hazard mitigation funding.

The next big obstacle they cited was that the digital divide and climate change are the result of inequitable access to information, the quality of that information, and the ability to interpret said information. A further concern was inequitable access to information quality and interpretation:

> There is an ever-widening gap between the haves and have-nots regarding the digital divide, both in expectations and access. Those well-connected expect personalized content and push alerts like large tech companies, while those without access are left behind as digital content is not tailored to their needs. Ensuring everyone has a WEA-enabled phone or weather radio is a start, but will not address fundamental inequities.

Other Challenges Identified by Emergency Managers

The survey respondents also identified these major obstacles to addressing the digital divide and climate vulnerability nexus:

- **Funding Challenges:** Difficulty securing sufficient, sustained resources for digital infrastructure in vulnerable areas.
- **Lack of Political Will and Leadership:** Weak political commitment to prioritizing digital equity within climate resilience planning.
- **Community Resistance:** Skepticism about new technologies or government-led initiatives in rural or underserved communities.
- **Educational Gaps:** Need for increased awareness about digital access in mitigating climate risks.

When these issues are a common occurrence, coupled with past experiences, it only heightens gaps in services and community mistrust, and the cycles repeat. These real-life insights from professionals on the ground reflect a critical awareness of how digital inequities align with and exacerbate climate vulnerabilities.

Connectivity Challenges in the Gulf South and Midwest

Following the discussion of the Gulf South, the Midwest region presents its unique challenges in the intersection of digital access, climate risk, and racial disparities. The vast rural expanses of the Midwest are particularly vulnerable due to the sparse population density, which often leads to economic disincentives for private investment in digital infrastructure. Urban areas, while better connected, still exhibit significant disparities in digital access among different racial and socioeconomic groups, compounding the challenges faced during extreme climate events.

In the rural Midwest, the digital divide is stark, with many areas lacking the basic broadband infrastructure necessary for timely communication and access to critical information. Environmental justice in these rural communities is not only about addressing pollution and land use but also about ensuring equitable access to digital technologies that can significantly enhance community resilience to environmental hazards. This lack of connectivity not only impedes emergency communications during events such as tornadoes and floods but also restricts access to telehealth services, remote education, and economic opportunities, further isolating these communities.

Urban areas in the Midwest face a slightly different set of challenges. Despite having better infrastructure, the digital divide remains pronounced along racial and economic lines. Research from the University of Chicago highlights significant Internet inequities that affect predominantly Black and Hispanic neighborhoods, where access to reliable and affordable Internet services is often limited (University of Chicago, 2022). This disparity affects these communities' ability to respond to and recover from climate-induced disasters, as well as their overall ability to participate in the digital economy and access government services.

The Department of Commerce offered programs aimed at addressing these disparities and providing excluded groups with access to more than just broadband, but also digital skills and literacy opportunities, telehealth services, and enhanced resilience measures. During the second Trump Administration, some federal grants were reorganized, reduced, or redirected, impacting consistency and scope of support for digital equity and climate adaptation (Center for Democracy & Technology, 2025; Congressional Research Service, 2025). Despite these shifts, targeted federal funding contributed to connectivity and community preparedness improvements in select areas. Stable, expanded federal support remains essential to sustain and scale gains.

Whether it comes from the federal government, state resources, or public–private partnerships, funding alone is insufficient. To address these issues, a comprehensive strategy that includes both investment in infrastructure and community-based initiatives is essential. For the rural Midwest, this could involve deploying mobile broadband solutions and satellite Internet technologies to reach the most isolated areas. In urban settings, efforts should focus on making high-speed Internet affordable and accessible to all residents, coupled with programs that enhance digital literacy and awareness of climate risks. Local governments and community organizations in the Midwest must work together to tailor climate resilience and digital inclusion strategies to the specific needs of their diverse communities (Willis, 2024).

Integrating community voices in these planning processes is crucial. By involving residents in the decision-making process, especially those from marginalized groups, policies and initiatives can be more effectively aligned with the community's needs and priorities. Participatory governance models can empower residents to share their unique insights and priorities, ensuring that local knowledge and experiences shape adaptation strategies. This participatory approach ensures that the benefits of digital inclusion and climate resilience efforts are equitably distributed, helping to break the cycles of inequality and vulnerability that have long plagued these regions.

Toward Digital Justice and Resilience

Recognizing the interconnectedness of the digital divide, climate risk, and resilience, it becomes evident that achieving digital justice is imperative for building resilient communities. Digital justice emphasizes the equitable distribution of digital resources, opportunities, and benefits across all segments of society. Digital justice expands beyond mere access to include the capability to use digital resources effectively for community empowerment and resilience (Sanders, 2024a, 2024b). Democratizing climate change information is an essential aspect of digital justice and resilience-building. Access to accurate and timely climate-related data enables communities to make informed decisions, develop effective adaptation strategies, and engage in meaningful climate action. However, the digital divide exacerbates information asymmetry, leaving marginalized groups at a severe disadvantage in accessing vital knowledge and resources. This lack of access to climate information perpetuates existing disparities, as those who are already marginalized are less equipped to respond and adapt to climate risks. Therefore, bridging the digital divide is vital for ensuring that climate information is accessible to all, facilitating more equitable and inclusive approaches to climate resilience.

A key component of digital justice is the democratization of climate change information. Access to accurate and timely data allows communities to make informed decisions, formulate robust adaptation strategies, and engage proactively in climate action. Currently, there exists a significant information asymmetry exacerbated by the digital divide, which leaves marginalized groups without crucial knowledge and resources. This disparity not only hampers their ability to respond to climate risks but also perpetuates existing social inequities.

Efforts to democratize climate data should focus on the development and dissemination of open-source platforms that provide real-time climate information accessible to all. This approach aligns with initiatives like the Open Disaster Data initiative, which advocates for transparent and accessible disaster-related data to enhance national resilience. Such platforms can empower communities, particularly those in vulnerable regions, by providing them with the tools needed to anticipate, respond to, and recover from climatic events.

Addressing Systemic Challenges and Advancing Digital Justice for Climate Resilience

To bridge the digital divide and mitigate its impact on climate risk, concerted efforts are required at various levels. At the policy level, there is a need for comprehensive strategies that prioritize equitable access to digital infrastructure and resources. Governments should invest in expanding a portfolio of connectivity options in underserved areas, particularly in rural and marginalized communities. One practical step involves creating community technology hubs that offer free or low-cost Internet access and digital literacy training. These hubs serve as critical spaces for vulnerable populations to develop the skills necessary to use digital tools effectively and participate in resilience efforts.

Policy frameworks must also evolve to recognize high-speed Internet as a fundamental right essential not only for everyday communication but for safety, civic participation, and emergency response. This is especially important in areas identified by our survey, where emergency managers noted significant gaps in the public's understanding of digital tools during crises. Climate education programs should also integrate digital tools and platforms to ensure that communities have access to the most current climate information. These programs can help communities better understand and prepare for the impacts of climate change, fostering a culture of proactive engagement and resilience building (Sanders, 2024a; Warschauer, 2003).

Regular impact assessments of digital divide interventions can help measure effectiveness and guide adaptive strategies to meet communities' evolving needs. The insights from emergency managers and resilience professionals underscore the urgency of inclusive policies that address both digital and climate inequities. Policymakers and stakeholders must prioritize investments in digital infrastructure as essential components of disaster preparedness, response, and recovery in regions most at risk.

Collaborative Efforts Across Sectors

Achieving meaningful progress in digital justice requires collaboration across the public, private, and nonprofit sectors. Partnerships among these sectors can facilitate the deployment of innovative solutions that address the specific needs of underserved communities. For instance, community-based broadband initiatives can be developed through public–private partnerships, leveraging private-sector technological expertise and public funding to create sustainable infrastructure.

Technology companies also have a critical role to play by developing affordable, user-friendly digital tools tailored to the needs of marginalized populations. These tools can simplify the process of accessing climate risk information and engaging with disaster response and recovery services.

Nonprofit organizations and community groups are vital in advocating for digital justice and providing essential services such as digital literacy training. These organizations ensure that individuals not only have access to digital tools but also possess the necessary skills and knowledge to navigate the digital landscape effectively.

Enhancing Community Engagement and Participation

Community engagement and participation are pivotal for building resilience and ensuring the effectiveness of digital justice initiatives. Engaged communities are empowered to advocate for their needs and exert substantial influence on the planning and implementation of digital infrastructure projects tailored to their specific circumstances. Regular community engagement helps to build a sense of ownership and responsibility among residents, crucial for the long-term success and maintenance of implemented solutions. Effective community engagement is fostered through sustained dialogues, interactive community workshops, and robust participatory governance models that encourage contributions from all community members, with a special emphasis on including those from marginalized groups. These interactions help bridge the gap between technology experts and community needs, ensuring that technological solutions, such as early warning systems and digital communication tools, are accessible and useful to everyone, particularly those who are most vulnerable to climate impacts. For instance, in the Gulf South, leveraging local insights about flood patterns and traditional coping mechanisms significantly enhances the effectiveness of climate adaptation strategies. This localized approach ensures that strategies and policies are not only technically sound but also resonate with and are readily adopted by the local population, thereby increasing their overall impact and sustainability.

Vision for the Future: Building a Digitally Just and Resilient Society

The future of climate resilience depends on a society where every individual has access to digital tools and information necessary to anticipate risks, participate in resilience planning, and mitigate climate impacts. Digital justice depends on equitable access to resources and the democratization of climate information. Open and accessible data platforms can empower communities by improving access to emergency warnings, public services, and opportunities for participation in the digital economy and civic life. Education and collaboration are central to this effort. Digital literacy programs combined with climate education help vulnerable populations engage confidently with technology and make informed decisions. Cross-sector partnerships among government agencies, technology companies, nonprofits, and local communities can support the development of sustainable and user-centered solutions. Sustained community engagement is equally important, as participatory governance and ongoing dialogue ensure that resilience strategies reflect local knowledge and needs.

Achieving this vision requires policies that integrate digital inclusion with climate resilience at every level of government and community. Investments in digital infrastructure must be viewed as foundational to disaster preparedness and response, not optional add-ons. The insights from surveys of emergency management professionals highlight the pressing need for inclusive approaches that address the intertwined challenges of digital inequity and climate vulnerability.

Conclusion

Building resilient communities amid climate change requires commitment to digital justice. The connection between the digital divide, disaster, and climate risk demands urgent attention. Co-location of public goods and bias in policy construction contribute to uneven digital resource distribution, exacerbating marginalized communities' vulnerabilities. Addressing the digital divide enhances community resilience, democratizes climate information, and promotes environmental justice.

Bridging the digital divide is not merely technological progress but a moral imperative to create an inclusive, sustainable future. Through comprehensive policies, partnerships, and education, we can advance digital justice and resilience, fostering a society where no one is left behind in climate challenges.

References

Acosta, R. J., Kishore, N., Irizarry, R. A., Buckee, C. O., et al. (2020). Estimating the post-hurricane migration of Puerto Rico using data from the social network Facebook. *Proceedings of the National Academy of Sciences of the United States of America, 117*(51), 32719–32727.

Bullard, R. D. (2000). *Dumping in Dixie: Race, class, and environmental quality* (3rd ed.). Westview Press.

Bullard, R. D., & Wright, B. (2012). *The wrong complexion for protection: How the government response to disaster endangers African American communities.* New York University Press.

CASS. (2014). *Social sustainability in the context of environmental changes.* [Journal/Report].

Center for Democracy & Technology. (2025, May 23). *Trump administration terminates Digital Equity Act programs; states should still pursue digital inclusion and equity.* https://cdt.org/

Congressional Research Service. (2025, October 17). *Broadband Equity, Access, and Deployment (BEAD) Program: Background and selected issues* (IF12908). https://crsreports.congress.gov/

Cooper, L. (2023, August 21). How the digital divide worsens the human impacts of climate change. Human-I-T. https://www.human-i-t.org/digital-divide-climate-change/

Fast Company. (2021). *Top global risks: climate change, digital divide, infectious diseases, and unemployment*. https://www.fastcompany.com

Fast Company. (2021). *Article identifying global risks including climate change, the digital divide, infectious diseases, and unemployment.* https://www.fastcompany.com/

Foster, S. R. (1998). *Justice from the ground up: Distributive inequities, grassroots resistance, and the transformative politics of the environmental justice movement*, 86 Cal. L. Rev. 775. https://ir.lawnet.fordham.edu/faculty_scholarship/295

Mohl, R. A. (2002). The interstates and the cities: Highways, housing, and the freeway revolts. *Poverty & Race Research Action Council.* https://www.prrac.org/

National Emergency Number Association. (n.d.). *9-1-1 statistics.* https://www.nena.org/page/911statistics

Nguyen, T., Lehmann, C., Funk, R. J., & others. (2023). *PLOS Climate article on using machine learning to predict Scope 3 emissions. PLOS Climate.* https://journals.plos.org/climate/

PNAS. (2018). *Technology and population displacement after Hurricane Maria in Puerto Rico.* https://www.pnas.org

Rothstein, R. (2017). *The color of law: A forgotten history of how our government segregated America.* Liveright.

Sanders, M. C. (2024a). *Increasing national resilience through an open disaster data initiative.* Federation of American Scientists. https://fas.org/publication/increasing-national-resilience-through-an-open-disaster-data-initiative/

Sanders, M. C. (2024b). *Building a digital justice framework.* Federation of American Scientists. https://fas.org/publication/building-a-digital-justice-framework-pef-monica-sanders/

U.S. Census Bureau. (2021, August 12). 2020 Census illuminates racial and ethnic composition of the country [Press release]. U.S. Department of Commerce.

Unwin, T. (2019). *Digital economies at global margins.* MIT Press.

University of Chicago. (2022, May 9). *Mapping the digital divide: Data reveals internet inequities across the country.* https://news.uchicago.edu/story/mapping-digital-divide-data-reveals-internet-inequities-across-country

Warschauer, M. (2003). *Technology and social inclusion: Rethinking the digital divide.* MIT Press.

Willis, C. (2024, April 2). *Federal approval of funds to bridge the digital divide and increase internet access. The Michigan Independent.* https://michiganindependent.com/economy/federal-approval-funds-digital-divide-internet/

Yamori, K., & Sugawara, R. (2021). Communication-related vulnerability to disasters. *International Journal of Disaster Risk Reduction, 62*, 102412. https://doi.org/10.1016/j.ijdrr.2021.102412

4

Global Patterns of Digital Inequity and Climate Risk

Contemporary scholars have extended the classical concept of a social contract beyond the nation-state, arguing that global risks such as climate change, economic inequality, and humanitarian crises require new forms of collective governance that resemble a global social contract. In particular, the planetary boundaries framework developed by Will Steffen and his colleagues argues that humanity must govern economic and technological development within ecological thresholds that allow societies to remain within a "safe operating space" for human development (Held, 2004; Pogge, 2002; Rawls, 1999; Steffen et al., 2015).

Why Global Digital Justice is Climate Justice

Now we get to delve more into the global dimensions of the digital divide and its intersection with climate justice. Just as climate justice recognizes the urgent need for equitable solutions to address the disproportionate impacts of climate change on vulnerable communities, digital justice is emerging as a pressing global concern. The convergence of these two issues calls for global attention and coordinated efforts to bridge

M. C. Sanders, *Climate and Code*, https://doi.org/10.1007/978-3-032-28032-9_4

the digital divide and promote equitable access to digital resources and climate resilience (Iberdrola, 2022; van Dijk, 2020).

Drawing parallels with the examination of discriminatory legacies within the United States, we extend our perspective to explore how the scars of colonialism shape vulnerability in the most climate-exposed countries worldwide. Historical patterns of exploitation, marginalization, and unequal access to resources have left lasting imprints, intensifying susceptibility to the adverse effects of climate change. Just as redlining produced invisible boundaries of exclusion in U.S. cities, colonial extraction and infrastructure development produced inherited geographies of access and neglect (Steyaert & Gould, 2009). By situating the digital divide within this broader historical context, we gain a deeper understanding of the systemic factors that contribute to the entwined challenges of digital inequity and climate vulnerability.

To substantiate this argument, we draw on map and survey data that provide empirical evidence of the global nature of these challenges. Southeast Asia, which is home to approximately 250 million people lacking Internet connectivity, exhibits heightened vulnerability to the impacts of climate change, including typhoons, floods, and sea-level rise (Krishnan et al., 2023; Shanahan & Bahia, 2023). Reports from UNDP, UNDRR, and GSMA illustrate how countries with low digital preparedness often overlap with those facing the most frequent and severe climate-related hazards (ITU, 2022; Mizutori et al., 2020). The correlation between limited digital access and increased climate risk is no coincidence; it demonstrates that digital exclusion is not simply a development gap but a compounding risk factor.

A multidimensional analysis reveals the systemic barriers that perpetuate the digital divide and hinder adaptation and mitigation efforts in marginalized communities worldwide (Gupta et al., 2023; World Bank Group, 2025). Understanding the global dimensions of these challenges empowers us to identify shared experiences, learn from successful interventions, and collaborate toward a more inclusive and sustainable future.

In the sections that follow, case studies and research findings from Southeast Asia, Latin America, Europe, and Africa highlight the diverse manifestations of the digital divide and climate vulnerability. These examples illustrate the interconnected nature of the issues and the urgent need for globally coordinated action.

The digital divide is not confined to any one region. It is a global phenomenon that intersects with climate justice. The historical legacies of colonialism, combined with contemporary disparities in digital access, exacerbate the vulnerability of communities to climate impacts. Bridging the digital divide and addressing climate vulnerability requires collaboration across governments, private actors, and communities, along with innovative approaches and a commitment to equity and social justice. By situating digital justice alongside climate justice, this chapter underscores that no resilience strategy is complete without both.

Historical Legacies: From Redlining to Colonial Extraction

To understand the global dimensions of the digital divide, we must first examine its historical roots in the United States. Redlining, a discriminatory practice that systematically denied access to housing loans and services based on race, caused significant economic harm to marginalized communities and perpetuated racial segregation. This deliberate exclusion had generational effects, shaping the socioeconomic landscape and contributing to disparities that persist to this day (Moore & Vitale, 2018).

Redlining emerged during the 1930s, when the Home Owners' Loan Corporation (HOLC) created maps to assess mortgage lending risk across U.S. cities. These maps were color-coded, with "hazardous" neighborhoods, often those with higher Black, immigrant, or working-class populations outlined in red. Banks, insurers, and government agencies used these classifications to deny mortgage access, resulting in systematic exclusion of entire communities from opportunities to build wealth through homeownership (NYC DOHMH, n.d.).

The effects of this practice were immediate and long-lasting. Black families, in particular, found themselves unable to access the same forms of subsidized loans and financial security available to white families moving to suburban areas. Egede et al. (2023) describe redlining as "the systematic denial of services such as mortgage access to residents of certain neighborhoods, primarily affecting communities of color, with enduring consequences even after the passage of the Civil Rights Act" (p. 1535). As property values stagnated or declined, so too did investment in public goods such as schools, hospitals, and parks.

Redlining was not limited to mortgages. It intersected with other discriminatory policies, such as racially restrictive covenants, exclusionary zoning ordinances, and predatory lending practices. These systems reinforced racial segregation and ensured that Black and immigrant communities remained concentrated in under-resourced areas. Lynch et al. (2021) note that "the legacy of structural racism is evident in current disparities in mortgage lending and health outcomes across communities with a history of redlining."

The intergenerational impacts are profound. Families who were excluded from federally backed mortgages missed out on the single largest mechanism for middle-class wealth accumulation in the twentieth century. This not only reduced household wealth but also created neighborhood effects that limited employment opportunities, educational attainment, and health outcomes for generations. The story of redlining, then, is not just one of discriminatory lending; it is a structural foundation for persistent racial inequities in the United States.

The economic disinvestment initiated by redlining extended into the 1970s and 1980s, a period marked by urban austerity, the decline of manufacturing, and widespread suburbanization. Cities responded to fiscal crises by cutting services, closing schools, and deferring infrastructure maintenance. For neighborhoods already weakened by redlining, this amounted to compounded marginalization. What had begun as financial exclusion in the mid-twentieth century became entrenched spatial inequality by the late twentieth century.

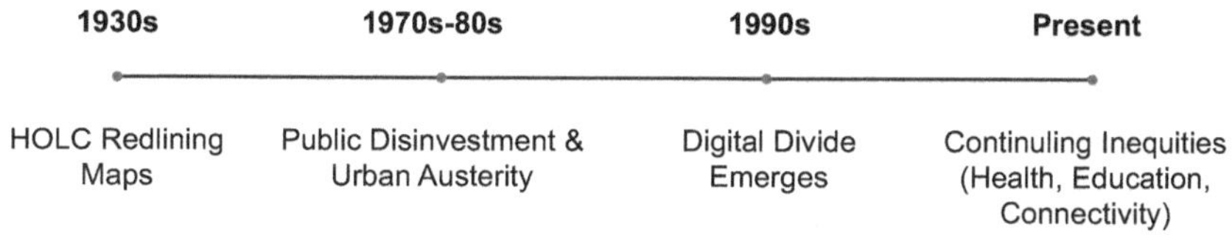

By the time the digital age began to accelerate in the 1990s, these communities were ill-positioned to benefit from new technologies. Broadband providers made cost–benefit decisions about where to build infrastructure, often skipping neighborhoods with lower incomes, aging housing stock, and weaker political clout.

In the early years of the digital age, a parallel form of discrimination emerged. "McCall et al. (2022) define this as digital redlining: the systematic disinvestment in broadband and digital infrastructure in low-income and minority neighborhoods." The term is used to describe how historically marginalized neighborhoods are systematically denied access to reliable, affordable Internet infrastructure. These inequities mirror patterns of disinvestment seen in housing and finance and extend their impact into disaster vulnerability: communities excluded from digital services lack access to early warnings, online recovery systems, and information necessary to adapt to climate change (Steyaert & Gould, 2009).

They argue that it is a modern analog to mortgage redlining, rooted in similar assumptions about risk and return. The consequences extend beyond Internet access. As telehealth, online education, and digital banking have become essential, communities subject to digital redlining face barriers to healthcare, economic mobility, and democratic participation. Wang et al. (2024) emphasize that "combating digital redlining is imperative for advancing health equity," noting that without intervention, these disparities will widen rather than shrink.

This exclusion from the information economy did not happen by chance. It was the logical outcome of decades of underinvestment in both physical and social infrastructure. The communities that were deprived of stable housing and financial services in the mid-twentieth century were the same ones deprived of broadband and digital opportunities half a century later. The cycle of exclusion shows how systemic inequities evolve rather than disappear, adapting to new technological eras while retaining old patterns of marginalization.

Colonial Empires, Infrastructure, and Development Outcomes

Globally, colonial extraction left similar scars. Just as redlining drew invisible lines of exclusion in U.S. cities, colonial infrastructures determined which regions were connected to trade routes, fiber optic cables, and later, modern digital systems, and which were left behind. This inherited geography of access is not accidental; it reflects centuries of prioritizing export economies and elite interests over local resilience.

While the U.S. story of redlining is rooted in domestic policy, similar patterns of structural exclusion appear in the legacies of colonial infrastructure investment. In both Africa and the Caribbean, colonial empires made decisions that privileged resource extraction and administrative efficiency over equitable development.

In Africa, both British and French colonial administrations concentrated investments in infrastructure such as roads, railways, and ports in coastal hubs and resource-rich regions. Ricart-Huguet (2021) finds that "colonial governments disproportionately invested in areas with natural harbors and historical trade hubs, while peripheral regions received little to no infrastructure." This uneven investment created durable development disparities, leaving inland and rural areas less connected to trade and, later, to global digitization.

Local African authorities sometimes shaped these dynamics. Bolt and Gardner (2020) show that native authorities influenced fiscal capacity through local taxation. The broader institutional structures ensured that these revenues reinforced existing hierarchies rather than redistributing opportunity. The end of colonial rule did not erase these inequalities; it simply left newly independent states with infrastructure designed to extract wealth rather than to connect citizens.

A similar story unfolded in the French Caribbean. Fadellin (2020) describes how colonial decisions around roads, electricity grids, and urban layouts produced uneven modernization that still constrains development today. She argues that "colonial infrastructures are not only physical but also cultural, producing enduring marks on the landscape that continue to shape economic and social life" (p. 4). The persistence of these infrastructures means that modernization and digitalization efforts often replicate, rather than overcome, the structural divides inherited from colonialism.

Taken together, these cases show that infrastructural inequities, whether rooted in redlining in the U.S. or colonial decisions in Africa and the Caribbean, have lasting consequences for who is connected, who is excluded, and how societies adapt to technological change. The tools of digitization, like mortgages and railroads before them, have been unevenly distributed along lines of race, geography, and history.

Mapping the Global Divide: Climate Risk Meets Digital Exclusion

Southeast Asia: Disconnected and Exposed

In global development, the phrase "where the road ends, development ends" underscores how infrastructure determines opportunity. In today's digitally interconnected world, the road is not only paved with asphalt but also with fiber optic cable and mobile towers. This reality is starkly evident in Southeast Asia, where approximately 250 million people remain offline despite the region's rapid economic growth and expanding urban centers (Shanahan & Bahia, 2023). Many of these disconnected communities live not in the relative safety of highland interiors but in floodplains, low-lying coastal areas, or cyclone-prone islands (Krishnan et al., 2023). The resilience of these communities is not only constrained by geography and climate; it is equally limited by their lack of access to digital infrastructure. Communities without reliable connectivity cannot access disaster alerts, emergency coordination platforms, online educational resources, or digital financial tools that enable recovery. In effect, the people who are most vulnerable to climate hazards are also those least connected to the digital tools that could help them prepare and adapt.

Colonial and Development Legacies

Much like in Africa and Latin America, colonial powers structured infrastructure systems for resource extraction and administrative control, not for inclusive development. Railways, ports, and telegraph lines in the late nineteenth and early twentieth centuries were designed to move raw materials from hinterlands to coastal export hubs, not to connect villages. Urban centers like Manila, Jakarta, and Hanoi were integrated into global trade networks, while rural and coastal populations were systematically excluded.

These patterns of uneven investment set the stage for the twenty-first century's uneven digital landscape. Today, metropolitan areas benefit from broadband rollouts, competitive mobile markets, and state-sponsored digital hubs. Meanwhile, rural, indigenous, and low-income communities remain disproportionately offline. Scholars have noted that digital access mirrors earlier colonial geographies of investment: connectivity flows to centers of power and profit, while peripheral regions are left behind (van Dijk, 2020). This continuity demonstrates how the legacies of colonial infrastructure decisions continue to shape vulnerability in the digital era.

Climate Exposure Meets Digital Exclusion

Typhoon Haiyan, which devastated the Philippines in 2013, revealed how communities without access to robust communication systems received little or no early warning. In coastal Vietnam, floodplain communities still rely heavily on word-of-mouth and radio alerts, leaving them less prepared for the increasing intensity of riverine floods. According to UNDRR's review of global disasters, Asia accounted for over 40 percent of global disasters between 2000 and 2019 and the largest share of people affected (Mizutori et al., 2020). Digital divides mean that warnings, aid coordination, and recovery programs do not reach those most in harm's way.

The UNDP's *Key Risks Report* further emphasizes that in the Asia–Pacific region, governance gaps, affordability barriers, and technological deficits limit the reach of resilience efforts (Krishnan et al., 2023). In practice, this means that the very regions hit hardest by climate hazards are the ones least likely to have affordable, reliable Internet access. Without deliberate policies to expand coverage and reduce costs, these communities remain locked out of life-saving systems.

A Paradox of Innovation and Exclusion

Despite these challenges, Southeast Asia is also a hub of digital innovation. Mobile penetration in countries such as the Philippines, Indonesia, and Thailand has expanded dramatically in recent years. Mobile banking and digital wallets have provided new pathways for financial inclusion, while social media platforms have become critical tools for disaster information sharing. Local innovations in early warning systems, from community-run SMS alert networks to open-source mapping projects, demonstrate how grassroots creativity can fill infrastructure gaps (Mizutori et al., 2020).

These advances are unevenly distributed. Wealthier households and urban residents disproportionately benefit from mobile platforms and broadband. Low-income users face high data costs and unstable network connections. In rural areas, where climate exposure is often greatest, mobile towers are sparse and coverage unreliable. The paradox is clear. Southeast Asia embodies both the cutting edge of digital innovation and the persistence of deep digital inequities. On the one hand, governments and private companies promote "smart city" initiatives, fintech solutions, and regional digital corridors. On the other hand, millions remain offline, unable to benefit from these advances. Reports from ASEAN and UN ESCAP emphasize that without deliberate investments in connectivity and affordability, the region's development and resilience goals will remain out of reach. The offline are not just excluded from convenience; they are excluded from survival tools (AlYahya, 2024).

The consequences are profound. Without connectivity, communities cannot access digital services that become critical during crises. They cannot participate fully in national adaptation planning or hold governments accountable through digital civic engagement. Digital exclusion thus functions as a climate risk multiplier; it compounds vulnerability, deepens inequality, and locks marginalized groups into cycles of risk. Unless deliberate investments in equitable connectivity are made, Southeast Asia's most at-risk communities will continue to be left behind, excluded from both development gains and resilience efforts.

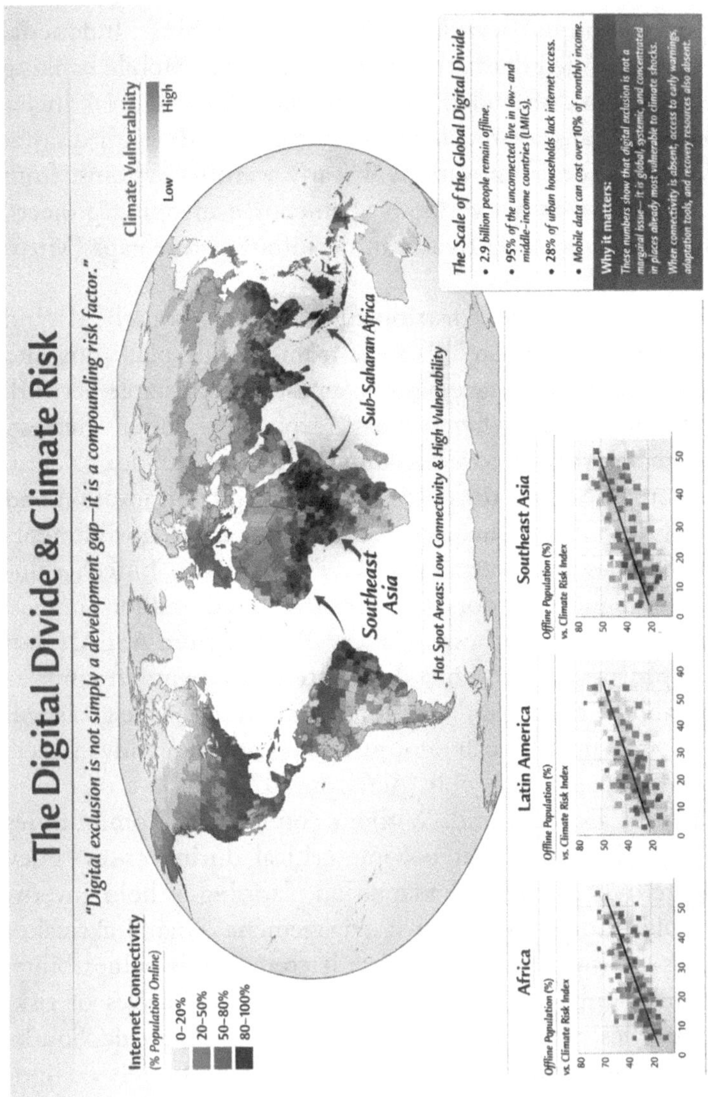
The Digital Divide & Climate Risk
"Digital exclusion is not simply a development gap—it is a compounding risk factor."
Internet Connectivity
(% Population Online)
0–20%
20–50%
50–80%
80–100%
Climate Vulnerability
Low
High
Southeast Asia
Sub-Saharan Africa
Hot Spot Areas: Low Connectivity & High Vulnerability
Africa
Offline Population (%) vs. Climate Risk Index
Latin America
Offline Population (%) vs. Climate Risk Index
Southeast Asia
Offline Population (%) vs. Climate Risk Index
The Scale of the Global Digital Divide
• 2.9 billion people remain offline.
• 95% of the unconnected live in low- and middle-income countries (LMICs).
• 28% of urban households lack internet access.
• Mobile data can cost over 10% of monthly income.
Why it matters:
These numbers show that digital exclusion is not a marginal issue—it is global, systemic, and concentrated in places already most vulnerable to climate shocks.
When connectivity is absent, access to early warnings, adaptation tools, and recovery resources also absent.
Sources: GSMA (2023), ITU, UN Reports

Latin America: Digital Inequality and Climate Adaptation Gaps

Latin America presents another illustration of the connection between lack of connectivity and climate risk. Across the region, millions of people remain excluded from Internet access (World Bank Group, 2025). This digital exclusion overlaps with multiple climate hazards, including drought in the Andes, deforestation in the Amazon, and hurricanes across the Caribbean and Central America (Mizutori et al., 2020). The communities most exposed to these threats are frequently those with the least access to digital systems that could help them adapt, survive, and recover.

Colonial Legacies of Infrastructure

As with other regions, the digital divide in Latin America cannot be understood apart from its colonial past. Infrastructure projects in the colonial and post-colonial eras were not designed for equitable development but rather to facilitate the extraction of resources and the consolidation of centralized power. In the Caribbean, for example, ports and communication systems were concentrated in coastal export hubs, reinforcing uneven access and economic dependency (Fadellin, 2020). This selective investment left rural and Indigenous communities without basic connectivity, a pattern that has carried over into the twenty-first century.

These colonial legacies created a template of exclusion that modern digital infrastructure often replicates. Broadband rollouts and fiber optic networks are concentrated in metropolitan areas such as São Paulo, Mexico City, and Santiago, while peripheral regions most exposed to climate hazards remain poorly connected. As Veiga and colleagues (2024) note, this uneven distribution of infrastructure has consequences not only for connectivity but also for how digital systems reinforce or challenge environmental justice.

Climate Exposure Meets Digital Exclusion

The overlap of offline populations and climate hazards is striking in Latin America. Rural Andean communities facing recurrent droughts are excluded from digital agricultural platforms that could provide weather forecasts or crop insurance. In the Amazon, Indigenous groups living at the frontlines of deforestation lack access to online tools that document and challenge land grabs or illegal logging. In the Caribbean, where hurricanes strike with increasing frequency and intensity, rural and low-income residents often cannot access early warning systems or disaster assistance portals.

UNDRR's *Human Cost of Disasters* report shows that the Americas experienced more than 1,200 disasters between 2000 and 2019, many of which disproportionately affected marginalized groups (Mizutori et al., 2020). At the same time, access to risk information and adaptation resources remains highly unequal. As the UN Economic Commission for Latin America and the Caribbean (ECLAC) has repeatedly stressed, gaps in infrastructure and affordability undermine adaptation capacity and exacerbate pre-existing inequalities. In effect, digital exclusion magnifies the vulnerabilities that geography and poverty already impose.

Innovation and Uneven Adoption

Despite these inequities, Latin America has also been a laboratory for digital innovation. Mobile banking platforms have expanded rapidly in countries such as Brazil, Mexico, and Colombia, offering new tools for financial inclusion. Civic tech initiatives have used digital mapping to document flooding in urban neighborhoods, while open data platforms have enabled citizens to monitor deforestation in near real time. These innovations demonstrate the potential for digital tools to transform resilience strategies.

Structural Inequities Reinforced

Digital exclusion in Latin America reinforces structural inequalities that extend across education, health, and political participation. During the COVID-19 pandemic, students in rural Mexico and the Dominican Republic were disproportionately excluded from remote learning, deepening existing educational disparities. Farmers in drought-prone regions of Central America lack access to digital climate advisories and crop modeling systems. Communities in Caribbean islands, already economically disadvantaged, struggle to navigate disaster aid systems that require online applications.

According to the World Bank Group (2025), affordability remains a critical barrier, with many households in the region unable to pay for reliable Internet service. Even where infrastructure exists, poor-quality connections and unstable electricity grids reduce its usefulness. Comparative work by Veiga et al. (2024) shows that underrepresented groups in both Brazil and Chicago face similar barriers to digital inclusion, underscoring that the digital divide is not only regional but structural and global.

These examples highlight the extent to which digital exclusion acts as a multiplier of vulnerability. It is not simply about whether households can stream entertainment or use social media; it is about whether they can access life-saving information, participate in democratic processes, and benefit from adaptation and recovery resources. The persistence of these gaps indicates that without deliberate intervention, climate change will widen the region's inequalities.

Toward Inclusive Resilience

Closing these gaps requires more than incremental broadband rollouts or one-off innovation pilots. Reports from CEPAL and ECLAC stress that meaningful digital inclusion requires systemic investment in infrastructure, affordability measures, and digital literacy programs. Equally important is ensuring that underrepresented voices are included in the design of resilience systems. Without such measures, even the most advanced digital tools risk reinforcing existing patterns of exclusion.

At the same time, Latin America's experiments with civic tech and environmental justice platforms suggest that alternative models are possible. If scaled equitably, these initiatives could create pathways toward resilience that do not replicate colonial logics of infrastructure investment but instead center marginalized communities. The challenge is ensuring that these innovations are not confined to urban elites but extended to the peripheries where climate risk is greatest.

Europe's Digital Conundrum and Estonia's Example

Although the digital divide is often associated with the Global South, Europe also grapples with distinctive challenges. The continent enjoys some of the highest broadband penetration rates in the world, with most households in Western and Northern Europe connected to reliable Internet service (ITU, 2022). In Eastern and Southern Europe, gaps in digital literacy, persistent socioeconomic disparities, and a lack of trust in government systems mean that significant portions of the population remain effectively excluded from digital participation. These divides illustrate that connectivity alone is insufficient: the governance and design of digital systems, as well as the willingness of people to use them, are equally important for building resilience.

Uneven Digital Inclusion

The European Union has invested heavily in expanding broadband and mobile coverage, aiming for near-universal access. But surveys reveal striking divides in digital literacy and trust. In Central and Eastern Europe, roughly one in four adults reports low digital skills, a figure that has barely improved over the last decade (ITU, 2022). Rural areas in countries such as Romania, Bulgaria, and parts of Greece lag not only in infrastructure quality but also in digital adoption, reflecting broader socioeconomic inequalities. Older adults across the continent are also more likely to be digitally excluded, even when affordable broadband is available.

These gaps manifest most clearly in the use of e-governance systems, which have become increasingly central to disaster preparedness, health-care delivery, and climate adaptation planning. Where citizens lack trust in government platforms, they are less likely to use them, leaving critical tools underutilized. In Italy and Spain, for example, surveys show that citizens are skeptical about the privacy and security of e-governance systems, reducing uptake despite technical availability. In Eastern Europe, bureaucratic inefficiencies and uneven service delivery further discourage participation. The divide is therefore less about whether cables reach households and more about whether people believe in and know how to use the systems to which those cables connect them.

Socioeconomic and Governance Barriers

Europe's digital divides are closely tied to broader socioeconomic inequalities. Regions with lower income, weaker educational outcomes, and higher unemployment often overlap with those that report lower digital participation. This pattern mirrors findings from global studies: infrastructure matters, but it interacts with poverty, governance, and social trust. The persistence of these divides in Europe, a region often held up as a digital leader, shows that exclusion is not limited to the Global South. It is a structural challenge that can persist even in high-income contexts.

Governance design also plays a key role. As Powell (2021) argues in her work on smart cities, the optimization of digital systems often privileges efficiency over equity. In European contexts, this has sometimes meant building e-governance platforms that function well for middle-class, digitally literate citizens but exclude those without skills, devices, or confidence in digital institutions. The result is a form of "soft exclusion," where services exist but large segments of the population cannot meaningfully access them.

Estonia as a Model of Digital Governance

Amidst this uneven landscape, Estonia has emerged as a model for digital governance. Following independence in the 1990s, Estonia invested in a comprehensive strategy of universal digital access, secure platforms, and streamlined government services. Citizens can vote online, access healthcare records, register businesses, and complete tax filings through a unified digital system. These investments have reduced bureaucratic barriers, enhanced citizen participation, and embedded resilience in governance processes.

Crucially, Estonia paired infrastructure expansion with efforts to build trust. Secure digital IDs, transparent governance practices, and public investment in literacy programs encouraged widespread adoption. By making digital services both accessible and reliable, Estonia demonstrated that exclusion is not inevitable. As Milan (2013) observed in her study of social movements and digital technologies, trust and usability are as important as technical capacity in ensuring participation. Estonia's approach reflects this principle by treating connectivity not as an end in itself but as a means to build inclusive civic engagement.

Estonia's experience highlights both the possibilities and limits of digital governance as a resilience strategy. On one hand, it shows that deliberate, inclusive strategies can bridge divides, even in contexts where resources are limited. On the other hand, Estonia is a relatively small and homogeneous country, raising questions about how transferable its model is to larger, more diverse European states. Even in Estonia, concerns remain about cybersecurity risks and the vulnerability of highly centralized systems to disruption.

The broader lesson is that digital exclusion in Europe is not primarily about a lack of infrastructure. It is about the interaction of governance design, public trust, and socioeconomic divides. Where trust is high and services are transparent, citizens adopt digital platforms readily. Where trust is low and bureaucracies are inefficient, exclusion persists regardless of technical availability.

Climate Risk and Digital Governance

These dynamics matter for climate resilience. Europe is experiencing increasingly severe hazards, such as heatwaves in Southern Europe, floods in Central Europe, and wildfires in Mediterranean states. Digital platforms play a growing role in coordinating response, disseminating warnings, and engaging citizens in adaptation planning. But if large portions of the population are unwilling or unable to use these platforms, the effectiveness of resilience efforts is compromised. As the ITU (2022) emphasizes, digital literacy and trust are as essential as broadband coverage for ensuring that digital systems support adaptation.

Estonia's example demonstrates that proactive governance can overcome these barriers. By embedding resilience into its digital strategy, it has created systems that enhance both efficiency and equity. Other European states can learn from this model by investing not only in infrastructure but also in digital literacy programs, transparent governance, and citizen trust-building initiatives. Without such measures, the gap between technical capacity and actual resilience will remain wide.

Africa: Innovation Amid Persistent Gaps

Africa presents both the starkest digital divides and some of the most innovative responses to them. Across the continent, nearly 60 percent of people remain offline, with the vast majority of the unconnected living in rural and low-income communities (GSMA, 2023a, 2023b; ITU, 2023). This lack of connectivity intersects with severe climate risks, including prolonged droughts in the Horn of Africa, recurrent flooding in West and Central Africa, and increasingly destructive cyclones in the south. As a result, many of the world's most climate-vulnerable communities also remain cut off from the very digital systems that could help them anticipate, adapt to, and recover from climate shocks.

Colonial Legacies and Infrastructure Patterns

Africa's uneven digital landscape is not accidental; it is rooted in patterns of infrastructure investment that date back to colonialism. As Bolt and Gardner (2020) and Ricart-Huguet (2021) have shown, during colonial administrations railways went to mines, ports were developed for export, and telegraph lines were rolled out to city centers for the convenience of colonial administrators. These projects bypassed rural areas and Indigenous communities, reinforcing a geography of neglect. The result was a continent in which infrastructure served external markets rather than local development.

That legacy persists today. Broadband expansion is concentrated in capitals and major cities, while vast rural regions remain poorly connected. Mobile penetration has grown rapidly, but unevenly, with affordability and infrastructure bottlenecks shaping who is connected and who is not. In many respects, the digital divide mirrors the colonial divide: connectivity follows lines of power and profit, leaving peripheral regions excluded.

Climate Exposure Meets Digital Exclusion

The consequences of exclusion are especially severe in a context of escalating climate risk. Drought across the Horn of Africa has left millions food insecure, yet many rural households lack access to digital advisories on rainfall, crop insurance, or humanitarian assistance. In West and Central Africa, floods increasingly devastate urban and peri-urban settlements, but without reliable mobile service or Internet, residents often do not receive timely warnings. Southern Africa's cyclone-prone regions face similar challenges: even when satellite data and early warning systems exist at the national level, the information cannot reach disconnected communities.

Data confirm these gaps. The GSMA's *Mobile Economy Sub-Saharan Africa* report highlights affordability as a central barrier: households spend a disproportionate share of income on data, making reliable connectivity inaccessible to the poorest (GSMA, 2023a, 2023b). The

ITU's *Africa Regional Digital Report* (2023) similarly emphasizes structural barriers, from the high cost of devices to the limited spread of broadband infrastructure. For households already grappling with climate-driven livelihood shocks, the added cost of connectivity places digital participation firmly out of reach.

Innovation Without Infrastructure

Despite these barriers, Africa has also become a hub of innovation in the absence of formal infrastructure. Perhaps the most well-known example is M-PESA, Kenya's mobile money platform that dramatically expanded financial inclusion. Without needing traditional banks or landline networks, M-PESA enabled millions to access remittances, savings, and credit. During climate disasters, mobile money has been a lifeline, allowing households to receive emergency funds or continue transactions when physical infrastructure collapses.

Other innovations have been equally striking. In Nigeria, community-driven mesh networks have emerged in urban peripheries where national broadband has failed to reach. These networks, built and maintained by local organizations, provide affordable connectivity that supports education and disaster communication. In Ghana, civic tech groups have developed open-source mapping and flood-monitoring platforms that integrate citizen data into resilience planning. These initiatives show that innovation does not always flow from national governments or multinational corporations; it often originates at the grassroots level, where necessity drives creativity.

Exclusion and Innovation in Tension

The contrast between exclusion and innovation highlights two realities. First, persistent digital inequity leaves climate-vulnerable populations disproportionately exposed to risk. When disasters strike, those without connectivity are the last to receive alerts, the least able to access aid, and the most excluded from recovery programs. Second, community-led digital infrastructures demonstrate that viable pathways to resilience exist even where formal systems lag. Local agencies have enabled populations to build networks of support that mitigate some of the risks of exclusion.

These grassroots systems are fragile. They often operate without long-term funding, supportive regulation, or integration into national resilience strategies. Without structural support, they risk being outpaced by the scale and intensity of climate disasters. Bridging the digital divide in Africa, therefore, requires more than connectivity alone; it demands a commitment to local empowerment, affordability, and resilience-oriented design.

Toward Inclusive Digital Futures

Policy interventions must recognize this dual reality. Expanding broadband and mobile infrastructure is essential, but affordability and literacy are equally critical. As the ITU (2023) and GSMA (2023a, 2023b) emphasize, high costs of data and devices create ongoing exclusion, even where networks exist. Addressing these barriers requires subsidies, regional cooperation, and investment models that prioritize social outcomes alongside profit.

At the same time, scaling grassroots innovations could redefine Africa's digital trajectory. Community-driven networks, mobile financial services, and civic tech platforms all show how resilience can be built from the bottom up. Rather than treating these as stopgap measures, governments and donors could recognize them as models for inclusive, climate-aligned infrastructure. Such recognition would ensure that innovation is not lost to neglect but becomes part of a broader effort to close Africa's digital divide.

This shift from digital exclusion to algorithmic exclusion represents the next frontier of inequity. Where digital redlining once determined who had access to broadband or affordable data, algorithmic redlining now shapes who receives loans, disaster alerts, or post-disaster assistance. Biased datasets and opaque decision-making systems can hardwire exclusion into climate adaptation tools, extending the reach of inequity into the future. Understanding this transition is critical: the patterns of digital inequity mapped across regions today are the foundation upon which tomorrow's algorithmic systems will operate. Without intervention, the climate-vulnerable may find themselves doubly marginalized, first by infrastructure gaps, and then by the very technologies intended to close them.

Algorithmic Redlining and Global Vulnerability Modeling

Disaster risk modeling and predictive analytics increasingly shape the global distribution of resources, from aid flows to insurance pricing to climate adaptation planning. Donors, insurers, and governments rely on models to decide which communities are "at risk," where investments should go, and which populations qualify for coverage. Yet the inputs to these models are uneven. Most rely on data generated in and about wealthier nations, excluding rural, Indigenous, and low-income communities in the Global South. As Silva and Kenney (2018) and Acemoglu (2021) argue, algorithmic frameworks reproduce structural biases when they are built on skewed data, encoding inequities into the very systems designed to reduce them.

The result is an "algorithmic divide," where the absence of data translates into the absence of aid. Quantitative evidence confirms this divide: Gupta et al. (2023) find that extreme weather events disproportionately disrupt Internet access for low-income and minority groups, reinforcing inequities precisely when connectivity is most needed. When these disruptions are not captured in models, affected communities become invisible to donors and policymakers. This invisibility is not accidental. It is produced by systems that equate measurable data with social reality, disregarding lives and experiences that fall outside of available datasets.

Structural Gaps in Global Models

Additional evidence reinforces the scope of this divide. The Insurance Development Forum (2022) documents persistent coverage gaps in climate insurance, leaving vast portions of the most climate-vulnerable populations without financial protection. The global insurance market is underpinned by probabilistic models that assume data availability, yet many Global South countries have weak or nonexistent hazard records, leaving risk assessments incomplete. The result is a paradox: those most vulnerable to climate shocks are deemed "uninsurable" because of insufficient data, perpetuating cycles of poverty and exclusion.

The IPCC's AR6 Working Group II *Cross-Chapter Paper on Data Gaps* (2022) highlights this same problem, emphasizing how data scarcity in the Global South undermines climate modeling and adaptation planning. The absence of high-quality meteorological data, demographic records, and socioeconomic statistics means that global models disproportionately reflect the realities of the North. As I argued in my 2023 Day One Project essay, this asymmetry means that digital inequity becomes embedded in the very algorithms meant to mitigate risk. Instead of producing equity, the tools themselves replicate and intensify injustice.

Bias in Modeling and Community Perceptions of the Divide

Much of the current discussion on digital divides and climate vulnerability is framed through top-down models and global indices. These tools provide essential benchmarks but risk overlooking how affected communities themselves perceive exclusion and resilience. Without these perspectives, modeling systems replicate biases and fail to capture lived realities.

The World Bank's *Voices of the Poor* project remains instructive. As one of the largest participatory poverty assessments ever conducted, it revealed how communities often experience exclusion not only as material deprivation but as social invisibility. Respondents described the absence of connectivity and information as a denial of dignity, a marker of not being seen or heard. These insights suggest that resilience cannot be measured solely in terms of broadband penetration or network coverage. It must also account for how people understand and trust the systems designed to serve them.

My own research underscores this point. In a 2021 talk at The Ohio State University's Translational Data Analytics Institute, I noted that "data gaps are not neutral—they determine which communities show up in risk models and which disappear." In Baltimore and Puerto Rico, for example, communities with the lowest levels of digital access were also the least visible in national disaster planning models. This pattern mirrors what we see globally: when populations lack connectivity, they fall out of the data streams upon which predictive analytics depend, rendering them invisible to systems of aid and governance.

Affordability, Trust, and Adoption

Recent ITU reports further complicate the picture. The *Measuring Digital Development: Facts and Figures 2022* report shows that even where infrastructure exists, affordability and trust remain the biggest barriers to meaningful adoption. Households across Eastern Europe, Latin America, and Africa report that broadband costs are prohibitive or that services are unreliable (ITU, 2022). Others express reluctance to use e-government platforms because of fears about surveillance, corruption, or misuse of personal data.

These perceptions matter for climate resilience. A government platform for flood alerts may technically "cover" a community, but if residents cannot afford data charges or distrust the state enough to avoid registering, they remain disconnected in practice. As I have argued in *The Regulatory Review* (2025), coverage does not equal access, and access does not equal justice. Metrics that treat connectivity as binary obscure the layered barriers of affordability, literacy, and trust. Without integrating these dimensions, models will systematically overestimate resilience in some regions and underestimate vulnerability in others.

Hidden Vulnerability and the Algorithmic Divide

The evidence from modeling systems points to the same conclusion as the regional case studies: inequity is reinforced when those most at risk are excluded from the data itself. Models built on Global North priorities, combined with persistent insurance gaps and weak data collection in the South, fail to capture the realities of frontline communities. Perceptions of affordability and trust show that even when infrastructure exists, people may still be effectively disconnected.

This disconnect has profound implications. Algorithms built on incomplete or biased data produce outputs that appear objective but are deeply partial. Lending models may deny credit to households in hazard-prone regions because their communities lack documented transaction histories. Disaster early warning systems may prioritize alerts in urban areas with dense sensor networks while undercounting risks in rural villages. Climate insurance models may deem entire regions uninsurable because hazard data is missing. In each case, the absence of data becomes the justification for exclusion.

Silva and Kenney (2018) describe this as the "feedback loop of bias," where the exclusion of vulnerable populations from data production results in models that reinforce their exclusion. Acemoglu (2021) extends this critique, arguing that automation and algorithmic decision-making exacerbate inequality unless actively corrected. These insights suggest that unless deliberate steps are taken, algorithmic systems will hardwire inequity into climate governance.

Toward Equitable Modeling

Integrating community perspectives and Global South data into climate and digital risk models is not only a matter of fairness but also of accuracy. Without these perspectives, bias persists, and the legitimacy of modeling systems erodes. The *Voices of the Poor* project and more recent initiatives in Latin America and Africa demonstrate that when communities co-produce data, the resulting models capture realities otherwise hidden from view (Veiga et al., 2024). Equally important is recognizing that technical fixes alone cannot solve structural inequity. Expanding sensor networks or building larger datasets will not close the divide unless affordability, literacy, and trust are addressed. As ITU surveys show, people remain effectively disconnected if they cannot afford data, distrust platforms, or lack the skills to interpret warnings. Models that fail to account for these barriers will continue to produce skewed results.

Bias as a Structural Risk

The algorithmic divide reveals that vulnerability is both produced and hidden by the very tools meant to measure it. Models that exclude Global South data and community perspectives underestimate risk in the regions most exposed to climate change. Insurance markets that declare communities "uninsurable" reproduce inequities under the guise of actuarial neutrality. Early warning systems that assume coverage equals access leave millions disconnected even when infrastructure is present.

Unless these biases are corrected, digital inequity will not simply persist; it will become embedded in the algorithms that govern resilience. The consequence is a twofold marginalization: first by infrastructure gaps, and then by the very technologies intended to close them. Without intervention, the algorithmic divide will deepen disparities in climate resilience, leaving the most vulnerable communities both unseen and unsupported in the age of predictive governance.

Bridging the Divide: What Justice Looks Like

Digital inclusion is essential for climate resilience. As the UN Secretary General's *Roadmap for Digital Cooperation* notes, "Digital inclusion is no longer a nice-to-have, it's the foundation for equitable climate resilience." Examples worldwide demonstrate that bridging the divide requires more than cables, towers, or devices. In Rwanda, mobile platforms were leveraged to extend banking, health, and agricultural services into rural communities. This enabled households to receive climate advisories, transfer funds during droughts, and access telemedicine in flood-isolated areas. In Brazil and Nigeria, community mesh networks empowered neighborhoods to self-manage connectivity, creating locally governed infrastructures that provided access where national systems failed. In Estonia, investments in secure, universal access and streamlined governance illustrate how digital trust can be built into national systems, reducing bureaucracy and embedding resilience. These cases show that when inclusion is designed into systems, resilience grows. But they also highlight that justice cannot be reduced to connectivity. Instead, it involves reshaping who holds power over information, infrastructure, and governance.

Digital Sovereignty and Community Control

Beyond connectivity, movements for digital sovereignty and community-driven governance expand the conversation. As Kukutai and Taylor (2016) document, Māori communities in New Zealand developed frameworks for Indigenous data sovereignty, asserting rights over how their data is collected, stored, and used. These frameworks ensure that climate and disaster information is not only available but also aligned with cultural values and community-defined priorities. Such sovereignty frameworks reject the assumption that "more data" always equals "better resilience," emphasizing instead that justice requires data governance that respects identity, culture, and autonomy.

Citizen-led digital cooperatives also challenge top-down models. As Stefania Milan (2013) observes, *"Technologies are never neutral; they embody particular visions of society."* Alison Powell (2021) similarly emphasizes that civic action in digital systems can represent a form of *"undoing optimization,"* where communities resist purely corporate or state-driven efficiency logics and instead prioritize equity and participation. These perspectives highlight that what lies *beyond the digital divide* is not only improved access but a fundamental reshaping of power and governance in digital systems.

Across these examples, a common thread emerges: resilience grows when digital systems are co-created with communities rather than imposed upon them. Rwanda's mobile health and finance platforms, Brazil and Nigeria's mesh networks, Estonia's digital governance, and Māori data sovereignty each show that digital justice is not only about extending access but also about reimagining governance. When communities are included as co-creators, digital systems can both extend access and transform power relations.

This reframing has direct implications for climate resilience. Communities with agency over their digital infrastructures are better positioned to ensure that warnings, aid, and adaptation strategies align with their lived realities. Rather than being invisible in data gaps or sidelined in algorithmic decisions, they become visible, empowered, and capable of shaping the tools that affect their survival.

Conclusion: A Shared Challenge, A Shared Mandate

The global phenomenon of the digital divide intersects with climate justice, shaping the resilience and vulnerability of communities worldwide. Historical legacies of redlining, colonial extraction, and systemic inequalities continue to influence who has access to digital systems and who remains excluded (Kukutai & Taylor, 2016; Steyaert & Gould, 2009). By examining case studies from Southeast Asia, Latin America, Europe, and Africa, this chapter has highlighted the multifaceted nature of the issue and its far-reaching consequences. Limited connectivity consistently compounds climate risk, reinforcing existing inequities and undermining adaptation and recovery capacity (Mizutori et al., 2020; Shanahan & Bahia, 2023).

As this chapter has shown, inequities are not only infrastructural but also algorithmic. Predictive models and climate insurance schemes often rely on incomplete or biased datasets, embedding Global North priorities while overlooking the experiences of communities in the Global South (Acemoglu & Restrepo, 2021; IDF, 2022; IPCC, 2022; Silva & Kenney, 2018). Without deliberate correction, the absence of data continues to mean the absence of aid. Incorporating community perceptions of access and trust (ITU, 2022) ensures that resilience strategies move beyond coverage statistics and toward measures that reflect lived realities.

Bridging the digital divide and advancing digital justice requires collaboration among governments, technology companies, nonprofit organizations, and local communities. Comprehensive policies and investments in infrastructure must prioritize affordability, literacy, and accessibility for the most climate-exposed populations (Gupta et al., 2023; World Bank Group, 2025). Just as importantly, initiatives must support community-led approaches, from Indigenous data sovereignty frameworks in New Zealand to neighborhood mesh networks in Brazil and Nigeria (Kukutai & Taylor, 2016; Milan, 2013). These examples show that resilience is not only about connectivity but also about empowerment and governance.

Recognizing the global dimensions of the digital divide alongside climate vulnerability allows us to work toward transformative solutions that promote equity, inclusivity, and sustainability. Through collective action and a shared commitment to both digital justice and climate justice, we can create a future where no community is left behind, where access to digital resources is universal, and where the impacts of climate change are addressed with human dignity and progress in the foreground. Chapter 4 shows how communities transform exclusion into power, turning digital and climate justice into resilience from the ground up.

References

Acemoglu, D., & Restrepo, P. (2021). *Tasks, automation, and the rise in U.S. wage inequality* (NBER Working Paper No. 28920). National Bureau of Economic Research.

AlYahya, D. (2024, September 17). Comment: Together, we can end the digital divide that disenfranchises 2.6 billion people. *Reuters*. https://www.reuters.com

Basham, B. (2023, September 27). The South's communication infrastructure can't withstand climate change. *Scalawag*. https://scalawagmagazine.org

Bolt, J., & Gardner, L. (2020). How Africans shaped British colonial institutions: Evidence from local taxation. *The Journal of Economic History, 80*(4), 1189–1221. https://doi.org/10.1017/S0022050720000371

Egede, L. E., Walker, R. J., Campbell, J. A., Linde, S., Hawks, L. C., & Burgess, K. M. (2023). Modern-day consequences of historic redlining: Finding a path forward. *Journal of General Internal Medicine, 38*(6), 1534–1537. https://doi.org/10.1007/s11606-023-08051-4

Fadellin, N. M. (2020). *Empty promises: Infrastructure and colonialism in contemporary Caribbean culture* (Doctoral dissertation). University of Wisconsin–Madison. https://www.academia.edu/43147589/Empty_Promises_Infrastructure_and_Colonialism_in_Contemporary_Caribbean_Culture

Fast Company. (2021, April 21). Climate change, the digital divide, infectious disease, and unemployment are top global risks. *Fast Company.* https://www.fastcompany.com

Held, D. (2004). *Global covenant: The social democratic alternative to the Washington consensus.* Polity Press.

GSMA. (2023). *The mobile economy: Sub-Saharan Africa 2023.* GSMA. https://www.gsma.com

GSMA. (2023). *The state of mobile internet connectivity 2023.* GSMA. https://www.gsma.com

Gupta, Y., Liu, Z., & Mostafavi, A. (2023, December 14). Digital divide in disasters: Investigating spatial and socioeconomic disparities in internet service disruptions during extreme weather events [Preprint]. *arXiv.* https://arxiv.org/abs/2312.08276

Insurance Development Forum (IDF). (2022). *Protecting our future: Closing the global protection gap in climate risk.* IDF. https://www.insdevforum.org

International Telecommunication Union (ITU). (2022). *Measuring digital development: Facts and figures 2022.* ITU. https://www.itu.int

International Telecommunication Union (ITU). (2023). *Africa regional digital report 2023.* ITU. https://www.itu.int

Intergovernmental Panel on Climate Change (IPCC). (2022). AR6 Working Group II: Cross-chapter paper 12 – Data gaps. In *Climate change 2022: Impacts, adaptation and vulnerability* (pp. 2323–2339). Cambridge University Press.

Iberdrola. (2022). What is the digital divide and how does it affect society? *Iberdrola.* https://www.iberdrola.com

Krishnan, A., Happ, S., Robele, S., & UNDP RBAP Strategic Foresight Network. (2023). *Anticipating risks and uncertainties for Asia and the Pacific: 2023 updated key risks report.* UNDP Regional Bureau for Asia and the Pacific.

Kukutai, T., & Taylor, J. (Eds.). (2016). *Indigenous data sovereignty: Toward an agenda.* ANU Press.

Lynch, E. E., Malcoe, L. H., Laurent, S. E., Richardson, J., Mitchell, B. C., & Meier, H. C. S. (2021). The legacy of structural racism: Associations between historic redlining, current mortgage lending, and health. *SSM – Population Health, 14,* 100793. https://doi.org/10.1016/j.ssmph.2021.100793

McCall, T., Asuzu, K., Oladele, C. R., Leung, T. I., & Wang, K. H. (2022). A socio-ecological approach to addressing digital redlining in the United States: A call to action for health equity. *Frontiers in Digital Health, 4*, Article 897250. https://doi.org/10.3389/fdgth.2022.897250

Milan, S. (2013). *Social movements and their technologies: Wiring social change.* Palgrave Macmillan.

Mizutori, M., Guha-Sapir, D., United Nations Office for Disaster Risk Reduction (UNDRR), & Centre for Research on the Epidemiology of Disasters (CRED). (2020). *The human cost of disasters: An overview of the last 20 years (2000–2019).* UNDRR.

Moore, R., & Vitale, D. (2018). The digital divide and educational equity: A look at students with very limited access to electronic devices at home. *ACT Research & Center for Equity in Learning.*

New York City Department of Health and Mental Hygiene. (n.d.). *A brief history of redlining.* https://a816-dohbesp.nyc.gov/IndicatorPublic/data-stories/redlining/

Pogge, T. (2002). *World poverty and human rights: Cosmopolitan responsibilities and reforms.* Polity Press.

Powell, A. (2021). *Undoing optimization: Civic action in smart cities.* Yale University Press.

Rawls, J. (1999). *The law of peoples.* Harvard University Press.

Ricart-Huguet, J. (2021). The origins of colonial investments in former British and French Africa. *British Journal of Political Science, 52*(2), 736–757. https://doi.org/10.1017/S0007123420000678

Shanahan, M., & Bahia, K. (2023). *The state of mobile internet connectivity 2023.* GSMA.

Silva, S., & Kenney, M. (2018). Algorithms, platforms, and ethnic bias: An integrative essay. *Phylon (The Clark Atlanta University Review of Race and Culture).* Available via UC Davis or SSRN.

Steffen, W., Richardson, K., Rockström, J., Cornell, S. E., Fetzer, I., Bennett, E. M., Biggs, R., Carpenter, S. R., de Vries, W., de Wit, C. A., Folke, C., Gerten, D., Heinke, J., Mace, G. M., Persson, L. M., Ramanathan,

V., Reyers, B., & Sörlin, S. (2015). Planetary boundaries: Guiding human development on a changing planet. *Science, 347*(6223), 1259855. https://doi.org/10.1126/science.1259855

Steyaert, J., & Gould, N. (2009). Social work and the changing face of the digital divide. *The British Journal of Social Work, 39*(4), 740–753. https://doi.org/10.1093/bjsw/bcp022

Veiga, C., Sharma, A., De Oliveira, D., Lage, M., & Miranda, F. (2024, October 6). Urban computing for climate and environmental justice: Early perspectives from two research initiatives [Preprint]. *arXiv.* https://arxiv.org/abs/2410.04572

van Dijk, J. A. G. M. (2020). *Closing the digital divide: The United Nations.* United Nations.

Wang, M. L., Gago, C. M., & Rodriguez, K. (2024). Digital redlining—The invisible structural determinant of health. *JAMA, 331*(15), 1267–1268. https://doi.org/10.1001/jama.2024.1628

World Bank Group. (2025). *Digital for climate.* World Bank.

5

Rethinking Value: Capital, Work, and Access in a Climate–Digital Economy

If the social contract defines the distribution of rights, responsibilities, and public goods within a society, it also implicitly determines how value is created and distributed. Reexamining value is therefore central to rebuilding the social contract for a climate-impacted and digitally mediated world. Decisions about investment and development shape which communities receive resources, which risks are mitigated, and which populations bear the costs of environmental disruption.

Beyond Money: Revaluing Justice in a Digital and Climate-Impacted World

It is important that we delve into the intricate relationship between money versus the pursuit of climate and digital justice. Monetary considerations play a crucial role in shaping our legal doctrines, business practices, and societal priorities. If we challenge traditional notions of value and explore the interconnectedness of what things cost and who bears them, we can foster a paradigm shift that prioritizes justice, equity,

M. C. Sanders, *Climate and Code*, https://doi.org/10.1007/978-3-032-28032-9_5

and sustainability. This process will help us better understand the shortcomings of prevailing approaches and move toward comprehensive and holistic strategies.

As we move forward in questioning the social contract, we should also examine each of these aspects, case studies, policy frameworks, and financial strategies that have the potential to shape our pursuit of climate and digital justice. With an understanding of that intersection, we can lay the groundwork for the call to establish a new social contract that upholds the principles of equity, inclusivity, and sustainability in an increasingly interconnected and digital world.

The pursuit of climate and digital justice cannot be divorced from certain economic realities. Money and value are foundational in shaping law, policy, and societal priorities. As I and others have written over the years, equity must be designed into systems, not appended as an afterthought (Sanders, 2021a). Similarly, Görgen and Ullrich (2022) caution that digitalization in climate strategies risks reinforcing hierarchies unless justice principles are applied.

This begins with the earlier notions of digital redlining and community divestment we covered earlier, but continues into the depths of how artificial intelligence, data-driven tools, and sometimes the data itself deepens these inequities. As Vera et al. (2019) explain, "'extractive logic' disconnects data from provenance … and whitewashes data to generate uncertainty," a dynamic that also permeates financial and digital infrastructures. When brought into the climate and resilience face, the consequences for frontline communities are even more dire. These concerns are echoed by West (2023), who warns that artificial intelligence may embed inequities in climate governance unless guardrails are introduced.

All of these concerns point to a need to begin detailing and implementing clear steps toward new frameworks. These can be legal, institutional, or civic in nature (Gerrard & Haverkamp, 2023), since good policy and even better law come from a variety of sources. That said, one common thread among policymakers and budget holders is that they always need a "number." So, let's take a closer look at how the money moves.

Biased Capital and Structural Access Barriers

In a capitalist society, access to capital remains one of the most significant determinants of opportunity. Historically, property ownership and access to land were the foundations of wealth and political influence. Today, the foundational descriptions of who gets to own property and build wealth may have changed, but some patterns persist. During the industrial revolution, in which industry and entrepreneurship redefined our economy, there was no shortage of gatekeeping. This mainly came in the form of social and political access tensions between "old" and "new" money (Gomez Betancourt & Vallet 2024). In the digital era, the new gatekeepers are algorithmic systems that determine creditworthiness, grant eligibility, and even disaster aid distribution. The consequences are profound: financial and digital infrastructures now reproduce the very inequities that previous generations fought to dismantle.

Digital systems were once imagined as tools of democratization, yet research shows how easily they replicate bias. Studies by Acemoglu (2021) and Silva and Kenney (2018) document how automation and algorithmic lending concentrate wealth among those who already possess technical skills and social capital. These biases are embedded deep within data inputs. Historic patterns of exclusion are encoded into models that appear neutral but operate with discriminatory effects. Algorithmic bias shapes who receives business loans, which neighborhoods are prioritized for broadband, and how federal recovery funds are allocated. A report by Fuster et al. (2023) confirms that credit-scoring algorithms amplify intersectional discrimination across race, gender, and parental status. Shah (2023) extends this critique to venture capital, revealing how algorithmic screening tools reproduce legacy inequalities in startup funding. The combined effect is a feedback loop of inequity, where those already excluded from financial systems are also shut out of the digital pathways that might have provided alternatives.

For many communities, this exclusion is compounded by other layers of identity. People with disabilities face inaccessibility in digital interfaces and automated financial tools. At the same time, LGBTQ+ individuals have reported that they encounter bias in systems designed around heteronormative family or housing structures. (National LGBTQ Task

Force 2022) These intersecting barriers underscore that algorithmic discrimination cannot be treated as a single-issue problem. It is intersectional because we can see that bias manifests along multiple, overlapping dimensions.

When working in Puerto Rico, both in research and post-disaster services in 2017, I found that the digital infrastructure needed to verify identity, document losses, or upload proof of residence was inconsistent or inaccessible for many families. I observed that survivors in lower-income, digitally underserved communities faced significant obstacles when accessing post-disaster aid. Many lacked stable broadband, home Internet service, or nearby computing access.

These barriers mattered. Households in low-income municipalities and in communities with limited broadband access waited longer for benefits, navigated more denials and mistakes due to reliance on paper-based applications, and faced higher burdens of proof than applicants from better-served areas.

In 2024 and 2025, my team from The Undivide Project returned to Puerto Rico to work on a project that mapped access to healthcare versus access to the Internet or digital services. Even years after Maria, children living in digitally divested and economically stressed areas continued to experience gaps in Medicaid access, delayed specialty care, and reduced eligibility follow-through (Sanders 2021b). These are not isolated administrative issues; they reflect how digital verification systems and eligibility rules amplify pre-existing inequalities.

Similar themes emerged in my study of disaster legal services. When examining legal services under the Federal Emergency Management Agency and other recovery programs, other scholars and I found that these digital and infrastructural divides align with deeper structural inequities. Even when aid was formally available, the administrative design assumed stable digital access, formal documentation readily at hand, and timely connectivity.

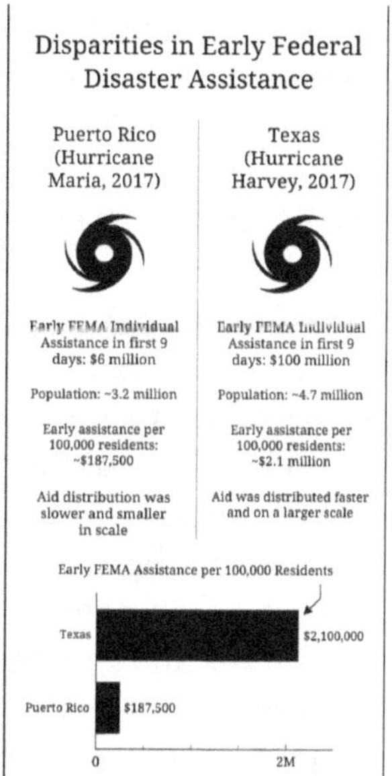

Comparison of early FEMA disaster assistance following Hurricanes Harvey and Maria. When normalized by population, early federal aid distributed to Texas communities was more than ten times greater than aid distributed to Puerto Rico in the same response window.

Source: FEMA disaster assistance data and federal disaster response analyses (2017 hurricane season).

Survivors from marginalized communities were more likely to need legal help to appeal FEMA decisions, correct identity mismatches, or challenge denials related to homeownership documentation.

FEMA assistance data provides a sharp illustration of this phenomenon. After Hurricanes Harvey, Maria, and Irma, Silva and Kenney (2018) found that applicants in affluent suburban Texas counties received higher median home repair grants than applicants in Puerto Rico, even when controlling for comparable levels of structural damage. The difference was not marginal; it stemmed from the cumulative advantage embedded in property valuation, insurance coverage, and documentation requirements. Homeowners in higher-income districts were more likely to have pre-disaster appraisals, clear titles, and access to digital submission portals. Puerto Rican families, by contrast, faced widespread title insecurity, lower pre-disaster valuations, and limited broadband. These barriers reduced not only the speed of recovery but also the amount of aid awarded. Two households with similar levels of damage effectively experienced two different systems of relief.

The View from Rural Communities

This insight shaped the Ten States Project in ways I did not anticipate when we began. Across the rural South and Midwest, community partners described digital exclusion as something that felt less like an inconvenience and more like a tax they paid every time a storm hit or a federal requirement came due. My team and I saw this firsthand. We met residents who drove forty minutes to the nearest public computer just to upload a FEMA document or check the status of an appeal. Local officials told us about evenings spent in small government buildings, waiting for slow systems to refresh so they could submit hazard-mitigation updates before deadlines closed. These were not isolated experiences. Small towns with limited staff explained that federal reporting, grant compliance, and routine planning diverted resources they could not spare. In those conversations, the pattern became clear: when federal systems demand high levels of digital fluency without ensuring the infrastructure exists to support it, the cost shifts downward to the very communities they are meant to help. That cost compounds with every storm cycle and every administrative cycle that follows.

The Iowa case study made this pattern even more visible. Local leaders, school administrators, and community organizations described how even modest gaps in connectivity changed what recovery looked like on the ground. Documenting storm damage that required photographs and uploads could take hours. Accessing energy-assistance programs stalled when online portals timed out or pages refused to load on older devices. My team noted how small differences in connection speed translated into tangible delays: reimbursement lags, rescheduled inspections, or missed opportunities to join statewide resilience initiatives because applications closed before the system cooperated. These challenges fell unevenly across the state. Communities with aging infrastructure, higher poverty rates, and limited technical capacity had the least bandwidth, both literal and administrative, to navigate complex systems.

Then when documentation was processed, flaws within the government systems drew out more inequity. A low-income ZIP code could signal "high risk," resulting in lower insurance and assistance payouts. In any of these communities, especially in rural contexts, a family lacking broadband access may miss documentation deadlines. An address flagged as "informal housing" a designation given to homes built by families or communities, or a barn with temporary housing, might not register in a damage-assessment model and might not be eligible for repair or replacement. Digital exclusion, in this context, was not simply an indicator. It was a mechanism that deepened socioeconomic disparities across counties.

The pattern across all of this work is consistent: the aid system's administrative design, particularly where it relies on digital tools or formal documentation, tends to reproduce the same inequities that disasters expose. These are design features of systems that privilege certain forms of data, expertise, and legitimacy. As Lee et al. (2022) observe, legitimacy itself depends not only on fair outcomes but on transparency and accountability in algorithmic processes. When the criteria for fairness are concealed, exclusion becomes self-perpetuating. The communities least likely to access capital are also least likely to understand how their exclusion is calculated or how to contest that exclusion.

Accountability as a Way Forward

So how do we fix this? One way is to think about who is designing these systems, what data they are using, and how community input is sought and managed. Legal scholars suggest pairing algorithmic-fairness mandates with fiscal-justice tools. Gerrard and Haverkamp (2023) propose integrating *equity audits* into public–private funding programs, requiring agencies to document how algorithmic credit and grant models affect underserved populations. Locally, community finance institutions could adopt open-data requirements and bias-testing standards similar to those in digital-rights legislation, ensuring transparency in climate and infrastructure funding. Locally, community finance institutions could adopt open-data requirements and bias-testing standards similar to those in digital-rights legislation, ensuring transparency in climate and infrastructure funding.

At the policy level, governments might think of requiring *impact statements*. Not an entirely new mandate, but something analogous to environmental impact assessments, which evaluate how emerging technologies affect equitable access to capital. Paired with stronger community oversight, these mechanisms can prevent algorithmic models from becoming the new redlines of economic opportunity. These findings reinforce the argument at the heart of this chapter: digital systems require focused audits, not just performance reviews. Efficiency metrics alone cannot capture the cumulative burden that digital exclusion places on under-resourced communities.

The Digital Wealth Gap and Generational Exclusion

Access to digital tools has also become a prerequisite for participation in economic, political, and civic life. Simply expanding broadband infrastructure or distributing devices does not guarantee equity. Digital inclusion requires the cultivation of digital literacy, citizenship, and trust. These are outcomes that are not automatically produced by access alone.

As Chan (2021) advocates, *"Internet access should be considered a human right, and the digital divide a social justice issue."* When framed this way, the digital divide becomes not merely a technical issue but a justice issue, an extension of civil rights into a new domain. The ability to navigate and use digital systems is as fundamental as the right to education or political participation. A recent IoT study demonstrates how gaps in digital literacy further reproduce inequities, preventing rural and other marginalized groups from participating fully in new economic ecosystems (Internet of Things Skills Study, 2020). Skills, capital, and access remain prerequisites for participation. Digital access requires more than hardware. The same research (IoT Skills Study, 2020) further reinforces this point: inequities in digital skills mirror broader socioeconomic and political divides. Those least likely to benefit from connected technologies are also those least represented in their design.

The results are particularly stark across generations. Younger people often adapt quickly to new technologies but remain underrepresented in governance or policy processes where digital infrastructure decisions are made. Older adults, by contrast, are frequently left out of digital transitions that affect their economic security or disaster preparedness. Studies of broadband adoption show that older residents in both rural and urban areas struggle to navigate online systems for insurance, financial relief, and health benefits, all of which leave them more vulnerable in the aftermath of climate-related events (ESCAP, 2021; UNDRR, 2019, 2022). Communities defined by disability, sexual orientation, and gender identity face additional structural barriers. Inaccessible interfaces can lock disabled individuals out of digital aid programs or online employment platforms. LGBTQ+ users may encounter data-collection systems that fail to recognize nonbinary identities or chosen families. (National LGBTQ Task Force 2022). When these overlooked communities are also overlooked in these systems, bias is essentially encoded into forms, categories, and eligibility requirements that appear neutral but operate unequally (Mello 2023).

In several parts of the United States, Indigenous communities experience these structural digital barriers in ways that are both acute and cumulative. During the 2022 Typhoon Merbok recovery in western Alaska, tribal leaders described a familiar pattern. Connectivity outages slowed damage assessments, hindered communication with federal and state agencies, and forced families to rely on radio and word of mouth to track supply deliveries. The storm's physical destruction was severe, but the digital gaps that shaped the recovery process revealed another layer of vulnerability. In communities where subsistence knowledge and kinship networks remain central to daily life, the absence of reliable digital infrastructure meant residents carried a dual burden: repairing homes and riverbanks while navigating emergency systems that were not designed with them in mind.

> For millennia, Indigenous Peoples have used their Traditional Knowledge to prepare for, cope with, and survive disasters. But Indigenous Peoples often do not have adequate opportunities to participate in the design,

> implementation, monitoring, and evaluation of state-led strategies and policies. UNDRR (2019, 2022)

These challenges recur across Indrough platforms that assume broadband access and stable cigenous territories in the Lower 48 states as well. In the Mountain West, tribal emergency managers have noted that wildfire alerts and evacuation maps are often delivered thellular coverage. During the 2020 and 2021 fire seasons, several communities adapted by combining local radio networks, text chains, and trusted messengers to fill information gaps. Those solutions reappear in Chapter 6 as examples of community-driven resilience that work precisely because they build on long-standing practices rather than replace them. The case study of Igloolik, Canada, provides a vivid example of how this adaptation unfolds in a different national and geographic context. The Iglulingmiut have faced decades of uneven investment, yet they have embraced information and communication technologies to strengthen emergency communication, expand economic opportunity, and sustain cultural life. Their experience previews how community-led innovation can take root even under extreme climatic and infrastructural constraints. What becomes clear in Alaska, the Mountain West, and Igloolik is that digital exclusion is not only an access issue. It reshapes the timeline, cost, and effectiveness of the response itself. Introducing these stories here prepares the ground for the next chapter's discussion of how communities design their own hybrid systems when federal technologies fall short, and why any new social contract must recognize those locally developed solutions as core public infrastructure rather than supplemental workarounds. For Indigenous and rural communities, digital exclusion compounds the risks of physical isolation, particularly in moments of crisis.

The experiences of Alaska, the Mountain West, and Igloolik also clarify a deeper point that carries forward into this chapter's next section: technology does not distribute power on its own. It reflects the conditions under which it is designed, funded, and governed. For Indigenous and rural communities, exclusion is rarely the product of a single absent device or broken tower. It emerges from a long history of underinvestment, jurisdictional fragmentation, and systems built around assumptions that do not match local realities. As a result, even well-intentioned tools can heighten inequity when they are deployed without attention to context. The gap between what a technology promises and what it delivers widens when the communities most affected have the least influence over how it is built or implemented.

This dynamic helps set the stage for the questions raised in the next section about emerging technologies. Communities already navigating digital exclusion understand that the harms of poorly designed tools are not abstract. They show up in emergency timelines, in eligibility rules, and in the invisible administrative costs that accumulate over the years. When new platforms, whether Web3, blockchain-based governance, or artificial intelligence applications, arrive with sweeping claims about decentralization or empowerment, those claims must be tested against the lived experience of communities that have seen similar promises before. The transition from economic and civic exclusion to the broader critique of emerging tech is not a shift in topic but an expansion of the same core inquiry: which technologies broaden access to opportunity, and which ones repackage old forms of exclusion in new technical language?

Which Tech Can Help? Which Ones Can Do More Harm?

This is a central question with which advocates and leaders should start when approaching each new advance in the rapid march toward a digitized global economy. One example of the need for this particular kind of query was with the advent of Web3, blockchain, and all things metaverse. The rise of AI, Web3, and blockchain technologies was initially heralded as a democratizing force, or at least a chance to bypass centralized institutions. In theory, decentralized autonomous organizations (DAOs) could allow communities to govern themselves digitally. In practice, they often replicated the same dynamics of exclusion seen in traditional systems. Access is still gated by technical literacy, capital, and connectivity. Fernandes, Müller, and Santos (2022) observe that DAOs in education and research have struggled to achieve diverse representation, while Marshall, Te Aho, and Ngata (2025) note that Indigenous-led blockchain initiatives face jurisdictional and governance challenges once external funding ends. This section critically assesses the claims of Web3 and regenerative finance, revisiting earlier interviews and examining how these tools can either reinforce or disrupt systemic inequities.

Can Decentralization Truly Democratize Access, or Does it Create New Hierarchies?

While blockchain systems promise open participation, the promise of decentralization often meets the reality of *digital elitism*. This refers to a system where those with coding expertise and financial resources dominate. In practice, decentralization without redistribution only scales inequity. As Etzioni & Etzioni (2017) observe, "autonomy without accountability becomes exclusion by design" and that "Regulation, not techno-libertarianism, may be the only path to ensure AI and blockchain systems serve human rather than institutional interests." True decentralization would mean devolving not just technical operations but governance and benefit-sharing. In a climate context, this could look like cooperatively managed community energy systems or tokenized carbon credits governed by local boards. Until these designs center accessibility, decentralized technology remains a digital echo of the inequalities it claims to disrupt.

These experiences illustrate that technology alone cannot redistribute power. Without intentional design, decentralization risks reproducing inequity at scale. Digital wealth and participation accrue to those already equipped to code, invest, or govern online systems. Communities historically excluded from physical institutions and infrastructures now risk exclusion from virtual ones as well.

Building Digital Capability as Resilience

Communities can operationalize inclusion by treating digital capacity as a measurable component of resilience. One approach is the development of *digital resilience scorecards*, which track broadband access, local IT training, and trust in online governance (Wood, 2024). Municipalities have begun experimenting with *justice-audit checklists* that pair physical-infrastructure metrics with digital-equity indicators (Gómez et al., 2024). Embedding these assessments into climate-adaptation or disaster-mitigation plans aligns technological investments with social outcomes. Over time, these tools could form the basis for a broader *Digital Equity Index*, or a mechanism for ensuring that every step toward modernization also advances justice.

In my own work, the Sustainability and Environmental Equity Dashboard (SEED) offers one applied model of how these ideas function in practice. By combining structured indicators, community-sourced narratives, and geospatial analysis, SEED tracks where digital capacity, environmental risk, and public investment diverge. It allows users to identify gaps, compare conditions across counties or neighborhoods, and follow how funding commitments translate into measurable outcomes over time. The Undivide Project team's hazard-mapping work reinforces this approach, showing that these metrics are most effective when they are built with local partners and tied to the decisions communities must make during mitigation and recovery.

These measurement tools also highlight a larger question: what counts as value, and who gets to define it? When communities develop their own indicators, they surface needs, histories, and priorities that rarely appear in standard cost–benefit analyses. They show how resilience depends not only on physical assets or digital access, but on the redistribution of decision-making power. That insight carries directly into the next section. Efforts to correct systemic inequities, whether through divestment, shareholder pressure, or public-sector commitments, cannot succeed if they rely on the same valuation models that created the disparities in the first place.

As digital-capacity tools evolve, they reveal more than access gaps. They expose the underlying assumptions about what counts as value in resilience planning. Those are: who defines it, who benefits from it, and whose needs remain outside dominant evaluation models. When communities generate and govern their own metrics, they can redefine value on their own terms. This reframing sets up the next question: if justice requires redistributing authority as well as resources, then divestment alone cannot achieve structural change. The challenge is not only where capital flows, but whether the systems that govern it are designed to shift power.

Divestment, Redistribution, and the Question of Value

Efforts to redress systemic inequities through divestment have become a defining feature of twenty-first-century advocacy. From fossil-fuel divestment campaigns to shareholder activism on social governance, the strategy has often been framed as a moral corrective: withdraw capital from harmful enterprises to realign markets with ethics. In practice, divestment has too often rewarded the institutions best equipped to perform compliance rather than those most willing to redistribute power. In both the climate and digital domains, divestment that is not paired with reinvestment in frontline communities risks replicating the very hierarchies it seeks to dismantle.

Many of us in academe and advocacy organizations have argued that justice is meaningful only when it yields measurable outcomes. Removing capital from extractive industries may signal virtue, but it does little for communities still living with the consequences of disinvestment. Similarly, ESG and climate-finance frameworks frequently label corporations as "sustainable" while excluding local stakeholders from decision-making or profit-sharing. The result is a financial ecosystem that prizes performative accountability over substantive change (Sanders, 2021a).

Vera et al. (2019) describe this dynamic as an "extractive logic." This is a pattern in which data and capital are separated from their social origins and repackaged as value for distant investors. Environmental data infrastructures, for example, can reproduce domination when the information collected from marginalized communities is used to monetize their vulnerability rather than mitigate it. "We aspire to build chemical inventories contextualized with community experiences and regulations that reflect the acts of environmental violence done to those living adjacent to chemical-producing facilities," Vera and colleagues write, underscoring that the burden of proof should rest with polluters, not residents. A more effective path forward requires shifting both resources and decision-making power. Mechanisms such as community-benefit agreements, participatory budgeting, and mutual-aid networks exemplify this shift. Singh, Martinez, and Tandon (2024) find that these models are most effective when intentionally designed to include disabled people, LGBTQ+ communities, elders, and youth. When inclusivity is embedded in governance, redistribution moves beyond charity toward shared authority.

These tensions reflect a broader pattern in both the climate and digital domains: systems often reward symbolic change over material redistribution. To understand why, scholars have traced how data, capital, and environmental risk are separated from the communities that generate them. Vera and colleagues describe this as an extractive logic. This is a cycle in which information and resources move upward while harms remain local. Recognizing this pattern makes clear why alternatives such as community-benefit agreements, participatory budgeting, and mutual-aid networks matter. They shift not only money but authority, anchoring governance in the places most affected by environmental and digital inequities.

Despite their promise, these models face persistent challenges. Funding cycles are short, evaluation metrics are inconsistent, and institutional partners often resist ceding control. Gerrard and Haverkamp (2023) note that operationalizing climate justice requires durable legal mechanisms—such as equity audits in grant programs and statutory requirements that a percentage of climate-adaptation funds flow directly to community-led projects. Without such frameworks, redistributive initiatives remain voluntary and easily reversed.

Throughout this book, recognizing the concept of divestment has been explored as a means to address systemic inequalities. The financial logic of divestment often rewards companies without shifting power.

ESG and climate-finance frameworks, for instance, may label corporations as sustainable while leaving frontline communities without resources or voice.

Even the most promising redistributive tools encounter institutional limits. Without legal mandates or durable accountability mechanisms, these initiatives rely on temporary funding or discretionary partnerships. Moving from aspiration to practice requires frameworks that can withstand political cycles and market pressures. This tension between vision and implementation leads directly to the next question: what costs do communities bear when redistribution fails to materialize?

> We aspire to build chemical inventories contextualized with community experiences and regulations that reflect the acts of environmental violence done to those living adjacent to chemical-producing facilities. This would shift the burden of proof from communities experiencing those harms to the responsible industries. (Vera et al., 2019) (EDGI)

Community-benefit agreements, equity audits, mutual-aid networks, and participatory budgeting offer models for redistribution rooted in justice. These mechanisms value care, community knowledge, and resilience over profit, reframing what counts as "value" in economic systems. Calls for divestment have gained traction, as have calls for outcomes-based approaches: justice is not about symbolic gestures but concrete value delivery (Sanders, 2021c). Shah (2023) shows that funding bias persists even in new "impact" markets, while Vera et al. (2019) argue that environmental data infrastructures often operate under extractive logics.

Adopting these steps transforms redistribution from aspiration into practice. As one community organizer in a flood-prone coastal parish explained, *"We don't need promises to pull money out of bad systems; we need investments that stay when the cameras leave."* The traditional financial logic of investment does undo disinvestment, and must therefore give way to a justice-based logic of reinvestment, or one that values endurance over symbolism and proximity over abstraction. By embedding accountability, inclusivity, and measurable outcomes into every dollar spent, we begin to reconstruct an economy in which value is defined not by who profits, but by who thrives. Work from Singh et al. offers a vision of redistribution that embeds justice in governance, rather than leaving it to corporate discretion (Singh et al., 2024). Some of their ideas include the following:

1. **Institutionalizing Community Reinvestment Targets.**
   Reserve a defined percentage of digital-infrastructure or climate-resilience funding for locally governed projects led by marginalized groups.
2. **Tying ESG Claims to Outcome Verification.**
   Corporations should substantiate sustainability ratings with independent audits measuring social as well as environmental impact.
3. **Rewarding Mutual Aid and Care Work.**
   Expand definitions of economic value to include informal labor that sustains communities during crises—childcare, elder support, and neighborhood response networks.

The True Costs of Preparedness and Participation

The pursuit of climate and digital justice carries multiple, overlapping costs. Those can be financial, social, and preparedness-related. Each reflects a different kind of inequity but converges on a common outcome: marginalized communities pay more, receive less, and are asked to prove their eligibility for protection. These "hidden" costs often go uncounted in conventional risk models, but they determine how quickly and how equitably a community can recover after a disaster.

Investments to bridge the digital divide or to fortify against climate hazards are frequently framed as expenditures rather than as rights. Yet the cost of exclusion is far greater. Reports from FEMA, the International Federation of Red Cross and Red Crescent Societies, and the United Nations Office for Disaster Risk Reduction (IFRC, 2014; UNDRR, 2022) document how communities of color, low-income families, and rural residents spend proportionally more to achieve the same safety thresholds as wealthier districts. For example, homeowners in flood-prone neighborhoods pay higher insurance premiums but receive slower assistance; families without broadband access must travel farther or rely on informal networks to apply for aid. These disparities do not end with infrastructure or insurance. As emergency management systems digitize, preparedness increasingly depends on models, algorithms, and data flows that can either narrow or widen gaps. When these systems are trained on incomplete or uneven data, they reproduce the disparities embedded in the landscape they are meant to correct. Zhang et al. (2025) find that AI-driven emergency response models, trained on incomplete data, often misallocate resources to already well-served regions. Automated systems designed to streamline decision-making can instead deepen disparities by prioritizing areas with stronger data coverage or political visibility. The 2019 UNDRR report underscores that the digital divide threatens to leave entire populations behind in early warning and response systems. Next Century Cities (2023) captures the real-world consequence: "Broadband scarcity amplifies environmental justice burdens through delayed aid, opaque planning, and exclusion."

> We had the water before we had the warning. — Iowa resident (NAS, 2021)

In rural states, the relationship between physical and digital exclusion is direct. During floods, hurricanes, and wildfires, communities without broadband connectivity often miss critical alerts or find themselves unable to upload documentation required for relief. Examples abound: the cost of building broadband access in rural Indigenous communities is often weighed against immediate infrastructure needs, leading to chronic underinvestment. Reports from FEMA and UNDRR highlight inequities in disaster preparedness costs, where marginalized communities spend proportionally more to meet the same safety thresholds (UNDRR, 2022).

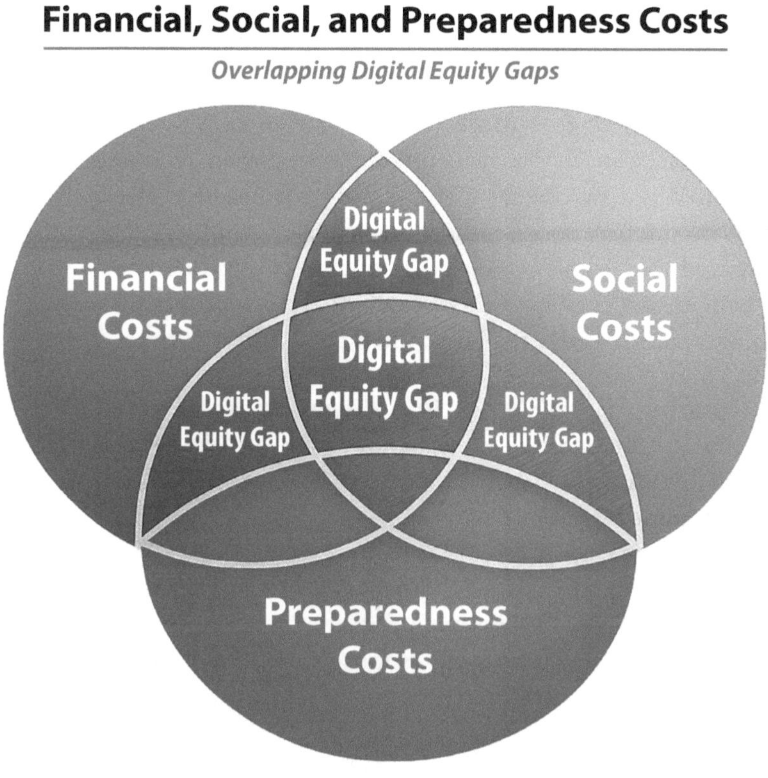

(Image created with Claude by author)

Social capital, meanwhile, functions as invisible infrastructure. The *Stronger Together* report by the National Academies (2021) emphasizes that communities with high trust and reciprocity recover faster after crises. When communication systems fail, these interpersonal networks of neighbors, local associations, and other informal caretakers fill the gap left by formal institutions. The financial logic of preparedness often privileges efficiency over equity. Cost–benefit analyses that drive federal mitigation investments tend to reward projects in higher-value areas, leaving low-income or rural regions with chronic underinvestment. In digital infrastructure planning, similar calculations prioritize urban centers over sparsely populated or Indigenous lands. The result is a self-reinforcing cycle: underinvestment leads to vulnerability, which in turn justifies continued neglect.

Breaking this cycle requires a recalibration of value. If resilience is defined as the capacity to adapt and recover, then preparedness should be measured not by dollars spent but by *who benefits*. The communities bearing the highest social and financial costs should also be those most empowered to design preparedness strategies. This demands a shift from cost–benefit to a framework that evaluates return on investment in terms of human security rather than profit. This could look like:

1. **Integrating Digital Access into Preparedness Funding.**
   Digital infrastructure—broadband, early warning systems, and digital literacy training—should be treated as critical resilience assets eligible for mitigation grants.
2. **Valuing Social Capital.**
   FEMA and municipal programs can incorporate *social connectedness indices* into preparedness planning, recognizing that trust networks and local knowledge reduce response time and recovery costs (NAS, 2021).

These steps would make resilience a matter of policy design and community engagement as opposed to chance. Preparedness must no longer be viewed as an individual responsibility or a discretionary budget line. The true cost of preparedness, then, lies not in the funds required to build resilient systems but in the moral and social debt incurred when we fail to do so. These costs also shape labor markets. The same structural factors that determine who receives timely warnings or equitable assistance also determine who can access stable employment, who can work remotely, and who has the digital skills needed in a changing economy. Preparedness and participation are thus tied to employment quality and digital citizenship. Understanding the future of work requires understanding these foundational inequities.

The Future of Work and Digital Citizenship

Preparedness is also conditioned by employment and social stability. In 2024, a pair of colleagues and I worked on research to understand how workplace benefits such as wages, benefits, and job security could be predictive of community resilience. Stable employment supports savings, social cohesion, and the ability to absorb shocks, while precarious work erodes all three. When disasters strike, financial insecurity becomes a multiplier for vulnerability (Wood et al., 2024). Employment is not only an economic variable; it is a social determinant of resilience. Growing gig economies, algorithmic management, and temporary contracts have eroded the social safety nets that once buffered workers from shocks. This becomes a more critical indicator in an environment meaning the concept of work as we have known and traditionally performed it has changed.

The meaning of work has always shifted with technology, but the speed and scope of transformation in the twenty-first century are unprecedented. Before the COVID-19 pandemic, conversations about the "future of work" centered on flexibility and automation. The pandemic made these theoretical debates immediate, exposing stark inequities in who could work remotely, who was labeled "essential," and who was left unprotected. For millions of essential workers, often from marginalized racial, gender, or geographic groups, safety became a privilege tied to digital access.

Acemoglu (2021), shows that automation shifts value away from routine labor, concentrating economic gains among those able to perform complex cognitive or technical tasks. This trend intersects with systemic inequities: the same communities excluded from quality education, broadband, and workforce training are those most vulnerable to automation-induced job loss. Sanders (2024) underscores that the U.S. federal job-classification system has failed to recognize emerging "green" and "digital" occupations that now form the backbone of sustainable development. This oversight limits investment in the very jobs, like community-based renewable energy installers, local data fellows, and resilience planners, that link environmental repair with digital inclusion.

Without explicit recognition in labor codes or workforce frameworks, workers performing climate-critical and digital-resilience roles remain invisible to policy. This invisibility has practical consequences: funding programs and workforce-training grants cannot support what they do not name. Modernizing these classifications is essential to expanding both economic opportunity and climate resilience—connecting labor justice to sustainable growth (Sanders 2024).

As digitalization accelerates, governance and policy have struggled to define what it means to be a "digital citizen." Access to remote work, online education, and civic participation now depends on the same infrastructures that determine resilience in disasters: connectivity, literacy, and trust in systems. ESCAP (2021) notes that digital inclusion is a precondition for sustainable development. Without deliberate investment in skills and accountability, the digital economy risks deepening inequality rather than expanding opportunity.

As these workforce transformations accelerate, they also reshape what it means to participate in civic life. Digital citizenship is no longer limited to accessing services; it encompasses the capacity to contribute knowledge, shape decisions, and negotiate fairness in algorithmic systems. It does not have to be a negative transformation. An alternative is expressed in this local project examined during a National Academies convening. "Youth-led mapping projects and school-based weather stations became trusted local data sources, turning students into climate communicators." (National Academies of Sciences [NAS], 2022, p. 66).

This example illustrates the transformative potential of digital citizenship. When young people gain access to technology and data, they become active participants in resilience-building rather than passive recipients of information. Digital literacy becomes both an economic and civic skill. West (2023) cautions that as artificial intelligence becomes embedded in climate and labor policy, fairness must be deliberately coded into its systems. Lee et al. (2022) further observe that perceptions of fairness depend as much on transparency and accountability as on outcomes. Efficiency without visibility erodes legitimacy; people must understand how algorithmic decisions about hiring, scheduling, or performance are made.

Looking ahead, the foresight literature signals that the convergence of automation, climate migration, and digitalization will influence where people can live and how they participate in the economy. Rodríguez and Khan (2025) describe divergent futures: one shaped by intensified precarity as automation and displacement accelerate, and another in which cooperative and decentralized ownership structures broaden worker autonomy. Borthwick et al. (2024) add that technology does not determine these outcomes on its own. Instead, the direction depends on whether policymakers invest in anticipatory planning and inclusive governance. This chapter turns to the question of how labor systems can evolve to meet the demands of a digital–climate era while protecting workers and communities.

Two ideas illustrate the type of reform required. First, job classifications need to be modernized so that emerging green and digital occupations are integrated into federal and international labor frameworks. Doing so would extend benefits, training opportunities, and labor protections to hybrid roles such as resilience data analysts, digital mappers, and clean-energy technicians. It also signals that these forms of work are central to climate adaptation and community resilience rather than peripheral or temporary. Second, governments and private institutions can strengthen foresight capacity by creating participatory scenario labs, where workers, communities, and policymakers test assumptions and co-design future-of-work pathways (Rodríguez and Khan 2025). Such forums help align climate planning with workforce inclusion and allow institutions to plan for transitions instead of reacting to crises.

Institutions, Infrastructure, and Reclaiming Digital Commons

Infrastructure defines the boundaries of participation. It shapes who can connect, who can decide, and who can profit. In the digital age, ownership of networks, data, and platforms has become as consequential as land ownership once was. To achieve a just transition, both digital and climatic, societies must confront not only inequities in access but the underlying question of *who owns the infrastructure of the future*.

The promise of Web3 and blockchain technologies has reignited debates about decentralization and collective ownership. Advocates argue that decentralized autonomous organizations (DAOs) and community blockchains can redistribute control and value. Yet emerging research suggests that these models are far from frictionless. Fernandes, Müller, and Santos (2022) note that while DAOs have the potential to democratize governance, they remain technically complex and often replicate exclusionary norms from traditional institutions. Marshall, Te Aho, and Ngata (2025) show that Indigenous-led blockchain governance faces challenges balancing sovereignty and interoperability with national legal frameworks. For communities already navigating historical marginalization, "decentralization" without inclusion risks becoming another form of displacement.

> Right now, almost everyone with a higher income has the devices and service they need to connect. But many lower-income people and people in more remote areas have none of that. — Next Century Cities (2023)

This disparity underscores a deeper structural issue: access without agency does not constitute justice. Ownership and governance, not mere connectivity, determine whether technology serves public or private ends. Chattanooga's publicly owned broadband network exemplifies what reclaiming the digital commons can look like. Lopez and Choi (2024) document how the city built a hybrid governance structure: municipal ownership with resident advisory councils empowered to guide reinvestment decisions. This model not only provided affordable high-speed Internet but also fostered local innovation ecosystems and civic participation.

At the same time, comparative studies reveal persistent regulatory and inclusion challenges. Municipal networks often face legal restrictions from state legislatures influenced by private telecom lobbies. Even successful community broadband initiatives struggle to sustain participation over time, as technical maintenance and governance fatigue erode engagement (Lopez & Choi, 2024). Indigenous DAOs encounter similar pressures—balancing culturally grounded decision-making with external funding cycles and jurisdictional constraints (Marshall et al., 2025).

These dynamics echo the regulatory barriers I have seen in grassroots broadband and data-equity efforts. In Baltimore, for example, community organizations working to expand neighborhood-level connectivity regularly confront permitting delays, fragmented ownership of utility poles, and inconsistent interpretations of municipal authority. Even modest attempts to build mesh networks or shared data hubs run up against rules written for large carriers rather than community-led networks. The result is a system in which residents can identify the solutions they need but must navigate structures that were never designed for small-scale or community-driven innovation.

The lesson from these cases is not that community control is unworkable but that it requires *infrastructure for governance* as much as governance for infrastructure. Digital commons must be supported by legal and institutional scaffolding that secures community rights, ensures equitable participation, and enables long-term sustainability. Görgen and Ullrich (2022) caution that without justice-centered governance, digitalization may entrench existing hierarchies rather than dismantle them. Similarly, Vera et al. (2019) show that data infrastructures reproduce extractive dynamics when designed without accountability to those whose information they exploit.

Promising counterexamples exist. Kumar et al. (2024) demonstrate that community broadband programs coupled with environmental data-sharing platforms enhance both resilience and participation. Next Century Cities (2023) finds that digital inclusion initiatives improve environmental governance outcomes when communities manage their own infrastructure and data. The UNDRR (2019) report further emphasizes that technology in disaster risk reduction must be "inclusive by design" and co-governed to prevent systemic exclusion.

A similar tension appears in policy transparency work. Through our writing with the Federation of American Scientists, my co-authors and I suggested open-data accountability frameworks to illustrate how public information can be structured as a shared resource rather than a controlled asset. We discussed open disaster data, but this can also be true of participatory dashboards, built with input from local partners, that show how communities can track funding promises, monitor implementation, and identify where commitments fall short. These tools function as a practical example of the digital commons in action: data is open, governance is shared, and accountability becomes a collective practice rather than a closed administrative process (Lakhina et al., 2023).

These ideas outline a broader shift toward treating digital and climate infrastructure as shared commons rather than market commodities. Municipalities and cooperatives can anchor this shift by adopting community ownership charters that define decision-making authority, representation, and revenue sharing, similar to Chattanooga's broadband model (Lopez & Choi, 2024). Governments can complement this approach by creating regulatory sandboxes that allow decentralized governance experiments, such as DAOs, to operate under adaptive rules while protecting communities from legal ambiguity (Marshall et al. 2025). Governance systems should also include periodic inclusion reviews to measure participation patterns and ensure that community-led digital or climate initiatives remain accountable and representative (Fernandes et al. 2022). These recommendations move digital-commons initiatives from theory toward institutional practice. They also challenge the assumption that "public" and "private" are the only governance models available. Between those poles lies a spectrum of community-based, cooperative, and hybrid structures that redefine what collective ownership means in a connected world.

The question is not whether communities can govern infrastructure but whether institutions will create the legal and financial space for them to do so. As one Indigenous technologist interviewed by Marshall et al. (2025) explained, *"Our sovereignty is not symbolic—it is coded."* That declaration reframes the debate entirely: the future of just infrastructure is not only technical but political, rooted in power, rights, and design.

Conclusion: Building a Social Contract for Equitable Value

The intersection of money, value, and justice in the pursuit of climate and digital justice demands critical attention. Biases in capital distribution, the challenges faced by marginalized communities in work and digital literacy, and the gaps in preparedness costs all reinforce systemic inequities. Yet examples of reinvestment, community ownership, and integrated adaptation strategies point toward a different path. By pairing qualitative community insight with quantitative justice indicators, policymakers and practitioners can evaluate progress toward an inclusive economy that measures success not by growth alone but by shared resilience and equitable participation.

Across this chapter, a clear pattern emerges: inequities in capital, digital access, work, and preparedness are not discrete failures but connected features of a system designed without justice at its core. When value is measured only in financial or technological terms, communities that sustain the social and environmental foundations of society remain excluded from both recognition and reward. A new social contract must redefine value to encompass resilience, care, and participation, and not solely profit.

Bias in the distribution of capital continues to shape who can innovate, adapt, and survive. Algorithmic systems now decide which neighborhoods receive infrastructure investment or whose disaster claims are processed first. These digital gatekeepers reproduce the same hierarchies once enforced through law and land ownership. I have argued that equity must be embedded into governance frameworks from the outset; only then can capital flows be rebalanced toward communities historically denied access (Sanders, 2021a). Gerrard and Haverkamp (2023) extend this view, urging policymakers to institutionalize fairness through equity audits and climate-justice impact statements.

The same principle applies to labor and digital participation. The future of work cannot be separated from the future of justice. Modernizing the way we classify work is a necessary step toward recognizing the labor that drives sustainability and resilience. The digital economy depends on the invisible work of educators, caregivers, and data stewards, yet traditional frameworks rarely account for their contribution. If these roles remain unacknowledged, inequality will continue to expand beneath the veneer of innovation.

The inequities revealed in preparedness systems are equally structural. Studies by UNDRR (2022) and IFRC (2014) show that marginalized communities bear the highest costs of adaptation and recovery while receiving the least protection. Preparedness, too, is a function of power:

who defines risk, who controls information, and who has the means to act on it. As my colleagues on the economic indicator research and I demonstrate, stable employment and local capacity strengthen resilience more than infrastructure alone. A social contract grounded in justice must therefore guarantee not only protection from hazard but participation in decision-making about what "safety" means (Wood et al., 2024).

Reclaiming digital and physical commons offers a path forward. The governance models emerging from Chattanooga's broadband initiative (Lopez & Choi, 2024) and Indigenous blockchain frameworks (Marshall et al., 2025) demonstrate that communities can govern infrastructure effectively when given authority and resources. These models reveal that inclusion is not a byproduct of technology—it is the precondition for its legitimacy. As Görgen and Ullrich (2022) warn, digitalization without justice risks replicating hierarchies; as Vera et al. (2019) add, data infrastructures that extract without accountability perpetuate environmental and social harm.

> The question is no longer whether technology can empower, but whether we will design power to be shared.

This chapter proposes that the design of power must begin with revaluing justice itself. Monetary and digital systems alike must account for care, social cohesion, and environmental repair as forms of value creation. Policy frameworks can advance this shift by pairing metrics of economic performance with measures of inclusion—digital access indices, resilience equity scores, and community governance indicators (Gómez et al., 2024; Sanders, 2024). These tools help translate justice from principle to practice, making it visible and measurable.

The call to action is not to invent new ideals but to restore old ones—to remember that public goods, from water to broadband, were once considered collective obligations. A social contract for equitable value builds on that legacy, expanding it to meet the realities of a digital and climate-impacted century. Justice, in this sense, is neither charity nor reform. It is design: a process of rewriting the systems that determine who counts, what counts, and how we measure progress. Linking back to Chapter 1's premise, justice is not an outcome but a design principle. To build a sustainable, resilient, and equitable future, we must embrace a social contract that redefines that future. It must center communities, mutual benefits, and resilience over the raw pursuit of profit.

References

Acemoglu, D. (2021). Tasks, automation, and the rise in U.S. *Wage inequality* (NBER working paper No. 28920). National Bureau of Economic Research.

Borthwick, S., Leung, A., & Patel, N. (2024). Anticipating AI-driven labor markets: Scenario methods for climate and digital work. *Futures, 158,* 103142. https://doi.org/10.1016/j.futures.2024.103142

Chan, C. (2021). The digital divide is a human rights issue: Advancing social work advocacy. *Journal of Human Rights and Social Work, 6*(2), 120–130. https://doi.org/10.1007/s41134-021-00163-w

Fernandes, D., Müller, A., & Santos, R. (2022). Web3 DAOs: An overview of development and possibilities for implementation in research and education. *Research and Education Technology Review, 12*(4), 233–247. https://www.researchgate.net/publication/361188860

Fuster, A., Goldsmith-Pinkham, P., Ramadorai, T., & Walther, A. (2023). Fair models in credit: Intersectional discrimination and the amplification of inequity. *arXiv preprint arXiv:2308.02680.* https://arxiv.org/abs/2308.02680

Gerrard, M. B., & Haverkamp, J. A. (2023). Operationalizing climate justice through law and policy. *Columbia Sabin Center for Climate Change Law.* https://scholarship.law.columbia.edu/sabin_climate_change/206/

Gomez Betancourt, R., & Vallet, G. (2024). The political economy of social change and nation-building during the progressive era. *Review of Political Economy, 36*(1), 1–7. https://doi.org/10.1080/09538259.2023.2279825

Gómez, E., Wang, J., & Porter, C. (2024). Justice metrics and community resilience: Evaluating digital inclusion tools for local governments. *Journal of Public Policy Analytics, 46*(2), 219–240. https://doi.org/10.1007/s11166-024-09440-8

Görgen, K., & Ullrich, S. (2022). Climate justice and digitalization. *Gaia, 31*(3), 146–150. https://www.oekom.de/_files_media/zeitschriften/artikel/GAIA_2022_03_146.pdf

International Federation of Red Cross and Red Crescent Societies. (2014). Effective law and regulation for disaster risk reduction: A multi-country report. International Federation of Red Cross and Red Crescent Societies. https://www.ifrc.org/document/effective-law-and-regulation-disaster-risk-reduction-multi-country-report

IoT Skills Study. (2020). Anything for anyone? A new digital divide in internet-of-things skills. *Policy & Internet, 12*(3), 330–354. https://doi.org/10.1002/poi3.171

Kumar, R., Singh, A., & Patel, S. (2024). Bridging the digital divide for effective environmental management. *Australasian Journal of Environmental Management, 31*(2), 145–160. https://doi.org/10.1080/14486563.2024.2373629

Lakhina, S. J., Gall, M., & Sanders, M. (2023). Increasing national resilience through an open disaster data initiative. *Federation of American Scientists.* https://fas.org/publication/increasing-national-resilience-through-an-open-disaster-data-initiative/

Lee, M. K., Jain, A., Cha, H. J., Ojha, S., Kusner, M. J., & Gajos, K. Z. (2022). Fairness perceptions of algorithmic decision-making: A systematic review. *Big Data & Society, 9*(3), 1–15. https://doi.org/10.1177/20539517221115189

Lopez, M., & Choi, Y. (2024). Governance in community broadband networks: Lessons from Chattanooga. *Telecommunications Policy, 48*(5), 102563. https://doi.org/10.1016/j.telpol.2025.102563

Marshall, J., Te Aho, L., & Ngata, H. (2025). Indigenous sovereignty and Web3: Balancing trust and technology in decentralized autonomous organizations. *Frontiers in Blockchain, 8*, 1630402. https://doi.org/10.3389/fbloc.2025.1630402

Mello, J. (2023). Digital accessibility: Disability justice and climate vulnerability in disaster response.

National LGBTQ Task Force (2022). Digital justice for queer communities: Technology, privacy, and safety.

Next Century Cities. (2023). Addressing digital divides is critical for achieving environmental justice. *Next Century Cities.* https://nextcenturycities.org/addressing-digital-divides-are-critical-for-achieving-environmental-justice/

Rodríguez, L., & Khan, Z. (2025). Climate migration, automation, and labor transitions: Foresight for inclusive economic planning. *Global Policy Futures, 15*(1), 44–61. https://doi.org/10.1007/s43681-024-00452-3

Sanders, M. C. (2021a). Moving beyond empowering the "vulnerable" to deliver value and outcomes in disaster legal services to the marginalized.

In D. H. K. Nguyen (Ed.), *Meeting the legal needs of disaster survivors: Third responders* (pp. 59–70). American Bar Association.

Sanders, M. (2021b). The confluence of poverty, the digital divide and disaster vulnerability on children: Baltimore and Puerto as case studies. Academia Letters. https://doi.org/10.20935/AL1478.

Sanders, M. C. (2021c). Equity in addressing climate change: Using law and policy to serve frontline communities. *Traumatology, 27*(1), 48–51. https://doi.org/10.1037/trm0000284

Sanders, M. C. (2024). Enhancing climate initiatives: Green jobs. *Federation of American Scientists.* https://fas.org/publication/enhancing-climate-initiatives-green-jobs/

Shah, P. (2023). Mitigating artificial intelligence bias in venture capital. *University of Miami Law Review, 78*(2), 215–248. https://lawreview.law.miami.edu/mitigating-artificial-intelligence-bias-in-venture-capital/

Silva, F., & Kenney, M. (2018). Algorithmic bias and access to capital: Digital redlining and exclusion in the platform economy. *Journal of Business and Society, 57*(4), 423–445.

Singh, A., Martinez, R., & Tandon, P. (2024). A systematic review of justice integration to climate resilience. *Environmental Science & Policy, 152*, 103629. https://doi.org/10.1016/j.envsci.2024.103629

Suresh, H., Guttag, J., & Barocas, S. (2023). Advancing algorithmic bias management capabilities in AI. *Information & Management, 60*(5), 103713. https://doi.org/10.1016/j.im.2023.103713

United Nations ESCAP. (2021). Digital development for inclusive growth. *United Nations Economic and Social Commission for Asia and the Pacific.* https://www.unescap.org/kp/2021/digital-development-inclusive-growth

United Nations Office for Disaster Risk Reduction. (2019). The use of technology in disaster risk reduction. *UNDRR.* https://www.undrr.org/publication/documents-and-publications/special-report-use-technology-disaster-risk-reduction

United Nations Office for Disaster Risk Reduction. (2022). Global assessment report on disaster risk reduction. *UNDRR.* https://www.undrr.org/gar2022

Vera, L. A., Walker, D., Murphy, M., Mansfield, B., Siad, L. M., & Ogden, J. (2019). When data justice and environmental justice meet: Formulating a response to extractive logic through environmental data justice. Environmental Data & *Governance Initiative (EDGI).* https://pmc.ncbi.nlm.nih.gov/articles/PMC6884364/

West, D. M. (2023). The US must balance climate justice challenges in the era of artificial intelligence. Brookings Institution. https://www.brookings.edu/articles/the-us-must-balance-climate-justice-challenges-in-the-era-of-artificial-intelligence/?utm_source=chatgpt.com

Wood, E. X., Lam, J. C., & Sanders, M. C. (2024). Identifying and assessing corporate employment variables that influence community resilience: A novel model. *Journal of Emergency Management, 22*(7), 27–38. https://doi.org/10.5055/jem.0808

Zhang, Y., Liu, H., & Wang, J. (2025). Implications of algorithmic bias in AI-driven emergency response. *Safety Science, 176*, 106095. https://doi.org/10.1016/j.ssci.2025.106095

6

Technology as Terrain: Navigating Power, Justice, and Resilience in the Digital Age

The opportunity to explore the digital terrain as both a frontier of opportunity and a site of contestation is critical to the conversation about climate and social justice. Building on the legal and policy frameworks discussed in the previous chapter, we investigate how emerging technologies, particularly artificial intelligence, blockchain, geospatial tools, and civic tech, can be leveraged to either reinforce historical inequities or transform systems toward regenerative and inclusive outcomes. We highlight where algorithmic bias and infrastructure gaps intersect with disaster risk and resilience planning, surfacing case studies and community voices that illustrate how digital justice must be designed into our climate solutions from the outset. Recent studies and community casework illuminate the urgent need for accountable, inclusive digital infrastructures.

> We are reminded that "resilience depends on networks of information, trust, and capacity as much as it does on physical infrastructure." If those networks are digitally unequal, so too will be our ability to adapt and recover (NASEM, 2021).

M. C. Sanders, *Climate and Code*, https://doi.org/10.1007/978-3-032-28032-9_6

From Legal Systems to Digital Systems

Law and policy alone do not shape the future. Technology, and the values embedded within it, are increasingly defining the boundaries of what is possible. As public services, disaster management, and climate mitigation become increasingly data-driven and tech-enabled, the question is no longer whether communities will interface with these systems, but rather how *and on whose terms.*

The digital realm is not just a set of tools, but a series of paths marked by uneven access, embedded bias, and contested control. Algorithmic decision-making in disaster response, satellite-based risk monitoring, decentralized finance tools, and digital public infrastructure all exist at the nexus of opportunity and exclusion. Without intentional design and governance, they risk reinforcing the very inequities they purport to solve. As Strate (2012) argues, *"If it's neutral, it's not technology."* The idea that digital systems are neutral tools obscures the social, political, and economic values embedded in their design. Borgmann (2012) adds that even the simplest technologies are expressions of power and intention, not just inert mechanisms. In the context of climate and disaster risk, these hidden dynamics become life-altering. Every artifact embodies choices about whose experiences matter. Strate (2012) argues that neutrality itself is a rhetorical device that "conceals intention behind convenience." When disaster-prediction systems privilege efficiency over equity, the resulting harm is structural, not accidental. Borgmann (2012) reminds us that tools "disclose worlds of power and purpose"; the digital world we inhabit discloses one where optimization often substitutes for justice.

The persistence of the neutrality myth benefits institutions that wish to appear objective. If technology is presumed to be impartial, inequitable outcomes can be treated as statistical noise rather than as a policy failure. Research by scholars such as Silva & Kenney (2018) on algorithmic racial bias and by Acemoglu (2007) on technological equilibrium shows that code reproduces context. True neutrality is impossible because any kind of design at scale is a social act. What is possible is *accountable partiality*, meaning we create systems that acknowledge their vantage point, publish their assumptions, and invite public oversight.

Algorithms and Injustice: When Tech Misses the Margins

Artificial intelligence and machine learning tools now underpin nearly every stage of disaster management, from early warning to recovery. Governments and private vendors deploy predictive models to estimate which neighborhoods are most at risk, how infrastructure might fail, and who should receive aid first. These tools promise efficiency: faster decisions, leaner bureaucracies, and quantifiable forecasts. Yet efficiency is not the same as equity. Algorithms built on incomplete or biased data routinely reproduce the hierarchies they claim to transcend.

The roots of these inequities are rarely visible within the technical systems themselves. They begin with the assumptions and data that feed computational models. Bernard (1987) termed this *machine design bias* the tendency of automated systems to mirror the worldview of their creators. When models rely on property values, tax records, or insurance claims as proxies for damage, they inherently privilege homeowners and formal economies. The outcome is predictable: disaster aid flows toward those who already have access to wealth and documentation, while renters, informal workers, and unregistered communities are undercounted or excluded. Algorithms, far from neutral, express a moral calculus about what is measurable and what matters. In the world they disclose, quantifiable loss carries more weight than lived experience.

Data itself is never entirely raw. Historical records used to train disaster models often omit the very populations most affected by systemic neglect, for example, rural residents, Indigenous nations, or Black communities in flood-prone areas. Acemoglu (2007) warned that "technological change tends to amplify pre-existing inequalities unless deliberately countered." In resilience planning, amplification manifests as compounding risk: the invisible becomes ineligible, and invisibility becomes policy. Because automation and other tools "inherits (sic) the hierarchies of its inventors," the same issues we see in the United States tend to track globally and in a variety of contexts (Bernard 1987).

Across other public systems, researchers have documented similar patterns: digital verification tools intended to streamline services inadvertently reproduce structural inequality. India had to navigate an identification system whose authentication failures blocked access to essential public benefits despite eligibility (Drèze et al., 2017; Khera, 2019). These failures arise not from individual misuse but from systemic design choices, unstable connectivity, inconsistent biometric capture, and opaque error resolution processes. Ramanathan (2021) describes this as *the architecture of exclusion*, noting that digital systems often "translate administrative discretion into computational denial." These cases reveal a broader truth relevant to AI-driven disaster tools: automated systems tend to encode existing hierarchies unless governance mechanisms explicitly counter them.

Bias replication occurs not through ill intent but through metrics. Machine learning systems optimize for performance indicators, such as speed, accuracy, and cost savings, but not for the less quantifiable indicators of justice. Once trained, they scale inequity with algorithmic precision. Disaster-response algorithms that value timeliness over inclusion inevitably classify marginalized populations as statistical outliers. To correct these distortions requires more than technical patches or better data; it demands a rethinking of governance and accountability. Participatory modeling offers one way forward. By including communities as co-designers and validators of local hazard data, these models can surface what institutions overlook: informal housing, social networks, or cultural sites of resilience. Ethical audits, algorithmic impact assessments, and open-data mandates are important steps, but they must be accompanied by shared control. Transparency without participation risks becoming another form of one-way oversight. Participatory governance can transform predictive analytics from tools of control into tools of equity.

Until equity is treated as a design requirement, not an afterthought, digital systems will continue to miss the margins they most need to serve. FEMA previously highlighted this need for collaboration, noting that "resilience is not a product but a process built through participation, partnership, and trust." The same principle applies to digital resilience. A just model does not aim for the near impossible task of eliminating bias, but to make the model's assumptions explicit and revisable. The goal is *accountable partiality*: a system that acknowledges where it stands and invites correction.

Civic Tech and Environmental Monitoring, or Who Gets to See, Speak, and Shape

Civic technology has emerged as one of the most dynamic and contested frontiers of environmental action. In the past, residents had to rely on government agencies or academic institutions to measure pollution or track flood exposure. Today, inexpensive sensors and open mapping tools allow communities to gather their own data. From grassroots air-quality monitoring in industrial corridors to participatory flood mapping in coastal neighborhoods, civic tech promises to democratize both knowledge and power.

That promise, however, depends on control. Projects that rely on volunteer data collection but store results on private servers or proprietary platforms risk turning civic participation into data extraction. Residents may generate measurements that improve institutional models but still lack the authority to interpret or challenge those results. Rossi (2018) writes, "Trust is produced through shared decision-making, not user interfaces." In other words, communities do not trust technology because it works; they trust it because they share in its governance. This is especially true when communities already distrust institutions due to historical neglect or environmental injustice.

Several ethical and procedural issues should and do arise in these situations. If communities collect data but cannot decide which determinants are used, it may not be participation at all, but an act of extraction. This question echoes long-standing debates in environmental justice, in which communities have fought to ensure that research conducted *with* them does not become research *on* them. The ability to collect data can expose pollution or flooding that official agencies overlook, yet if that data is harvested without local consent, the harm merely changes form. Examples of community-led governance show a different path. In California's Central Valley, residents have used open-source air monitors to track particulate matter and negotiate mitigation measures directly with local officials. In Puerto Rico, neighborhood mapping teams have documented stormwater drainage failures missed by federal surveys. These projects succeed not because of sophisticated tools, but because of shared authority. In these examples, data was hosted locally, interpreted collectively, and linked to action. They exemplify what Rossi (2018) calls *co-governance*: building trust not as a side effect of transparency but as an intentional architecture.

Still, not all initiatives follow this model. Many civic tech projects, particularly those driven externally on short-term grants, lack continuity once the funding ends. Others treat communities as data sources rather than as partners, offering dashboards but not decision-making authority. These models often reinforce dependence on outside experts rather than local knowledge. Civic technology is most powerful when it transforms participation into governance. That shift requires transparent systems, sustained funding, and genuine co-ownership of both processes. Such sustained partnerships can turn civic tech into a form of civic engagement with enduring and future benefits for everyone.

A justice-centered approach to civic technology recognizes data sovereignty as a foundation, not an afterthought. Indigenous data frameworks, such as the CARE Principles, "Collective Benefit, Authority to Control, Responsibility, and Ethics," illustrate how information can be governed in alignment with community values. Similarly, neighborhood data trusts or cooperatively managed mapping platforms ensure that those who collect data share in the interpretation, use, and benefits of the data. As Carroll et al. (2020) write, "Data is not a resource to be mined, but a relationship to be managed responsibly." As communities take control of the data that defines them, civic technology becomes what it was always meant to be: a tool for visibility, voice, and shared power.

AI and the Illusion of Inclusion

Artificial intelligence has become the defining infrastructure of the digital age. It analyzes satellite images to monitor deforestation, predicts flood risk, optimizes power grids, and drives new models for climate forecasting. In principle, AI could help accelerate resilience. Like every technology before it, AI reflects the conditions of its creation. The question is not whether AI will influence the future of climate and disaster response, but how it will shape the future of the system.

The United Nations Framework Convention on Climate Change (UNFCCC, 2024) highlights that "AI offers transformative potential for climate action, but also risks deepening global inequities if developing countries are excluded from access and governance." Access to computational resources, technical literacy, and high-quality data already skews toward wealthier states and corporations. When predictive models depend on data from the Global North, the Global South becomes a statistical afterthought. In climate adaptation, this means that the very regions most at risk often appear least visible in the algorithms guiding global policy.

These structural asymmetries mirror the inequities observed in earlier digital divides worldwide. The OECD (2024) cautions that without deliberate intervention, AI "may exacerbate existing inequalities by concentrating expertise and decision-making in high-resource contexts." In education, disaster management, and health systems alike, automation tends to reward those already equipped to interpret and apply its outputs. The same pattern holds for resilience planning: local officials with limited digital capacity may depend on proprietary AI tools whose assumptions they cannot audit or alter. This dependence shifts authority from the public sphere to algorithmic intermediaries. Borgmann (2012) described technology as a force that "discloses a world of power and purpose." AI's purpose, however, is not intrinsic—it is engineered through choices about data, design, and governance. When those choices remain opaque, accountability recedes. The result is a widening gap between those who generate risk and those who bear it, mediated by code that few can question.

AI's role in climate and health applications underscores this tension. Algorithmic systems used in health and environmental monitoring "perform best on populations that most resemble their training data," a limitation that "risks excluding the marginalized twice, first from the data, then from the benefits" (Susić et al. 2024). Models that rely on Eurocentric datasets or commercial satellite imagery may miss rural or informal settlements where hazard exposure is highest. In such cases, AI does not simply mismeasure reality; it reproduces inequality through omission.

Its potential for inclusion is equally powerful when grounded in justice-oriented design. The AI and Inclusion Institute (2024) emphasizes that "embedding local knowledge into algorithmic governance can transform AI from an extractive to a restorative force." These models echo the principles of civic tech discussed earlier: data sovereignty, transparency, and co-ownership. The same mechanisms that empower citizens in mapping pollution or floods can inform how we train and deploy intelligent systems.

The effectiveness of AI systems also depends on intermediaries that help communities navigate them. In Latin America, public libraries have played this role for decades, offering digital literacy programs and serving as trusted sites where residents can learn to interpret government or environmental data (Gómez, 2020). These institutions illustrate how AI-enabled platforms require not only technical accuracy but also the social infrastructure of trusted actors who can contextualize outputs and support informed use. Similar dynamics appear in Indigenous communities across Aotearoa New Zealand, where Māori data governance organizations have insisted that any use of AI involving Māori data must align with Indigenous values and cultural protocols (Hudson et al., 2020). These examples highlight a recurring theme: equitable AI requires intermediaries, governance structures, and cultural frameworks that ensure that communities lead in defining what digital information means to them, how they use it, and its context.

A justice-centered approach to artificial intelligence requires three elements: equitable access, participatory governance, and regulatory accountability. Equitable access ensures that developing nations and low-income communities can use, adapt, and audit AI tools relevant to their needs. Participatory governance invites community oversight, ensuring that those most affected by climate or economic shifts can influence design parameters. Regulatory accountability establishes ethical baselines that safeguard public trust. These are transparency, data protection, and redress mechanisms. Etzioni and Etzioni (2017) argued that "regulation, not techno-libertarianism, may be the only path to ensure AI systems serve human rather than institutional interests." The same holds today. Governance, not innovation, determines whether AI becomes a bridge or a barrier. AI for climate resilience must follow the same path as civic technology, with shared ownership: open data validated by communities, transparent models subject to public scrutiny, and benefit-sharing frameworks that return value to those who generate the knowledge. When intelligence is distributed, AI becomes less a tool of prediction for predication's sake and more a process of participation.

AI may process data at unprecedented speed, but infrastructure determines whether that insight reaches those most in need. Justice in the digital age will depend on how well we align the intelligence of machines with the realities of public infrastructures and the lived experiences of the people who depend on them. The next challenge is therefore structural: even the most inclusive AI models cannot achieve equitable outcomes if the public infrastructure around them is fragmented, privatized, or inaccessible.

Digital Public Infrastructure: Who Builds it Versus Who it Serves

Similar to civic infrastructure, public infrastructure is a critical component of this examination of climate impacts, digital justice, and community benefits. Public infrastructure once meant concrete and steel, the roads, bridges, and power lines that made modern life possible. Today, it increasingly includes invisible architectures of connection: broadband networks, cloud storage, data platforms, and identification systems that underpin nearly every public function. From disaster alerts and benefit applications to telemedicine and education, digital infrastructure is now as essential as electricity or clean water. Like physical infrastructure, the systems that carry digital life are often fragmented, privatized, or unevenly distributed.

This phenomenon is called data-deserting: "When data is missing, the most vulnerable disappear twice: first from services, then from solutions." When communities lack reliable access, they are invisible to the systems designed to protect them (Silva and Kenney 2018). Connectivity, once a convenience, has become a precondition for survival during a crisis.

Many early warning systems often presume consistent smartphone ownership and the ability to receive push notifications. In the United

States, emergency management dashboards assume English proficiency, stable Internet connections, and familiarity with geographic information systems. Platforms that deliver recovery resources rely on assumptions about documentation, digital literacy, and uninterrupted access. Those who fall outside these assumptions, whether because of age, disability, language, income, or geography, fall through unseen cracks in digital governance. The National Academies of Sciences, Engineering, and Medicine (2021) observed that "resilience depends on networks of information, trust, and capacity as much as on physical infrastructure." When those information networks exclude the most vulnerable, resilience remains a promise unfulfilled.

This is also true when daily government functions are digitized and privatized. Many public agencies now rely on corporate vendors to manage everything from digital IDs to benefits and procurement platforms. Contracts negotiated for efficiency often sacrifice equity, locking public functions into proprietary systems. When funding lapses or licenses change, continuity collapses, leaving citizens stranded. This phenomenon, or what Borgmann (2012) might call "the technological surrender of public purpose," reframes governance as a customer-service model, in which accountability is mediated through terms of service rather than through democratic oversight.

Even when digital infrastructure exists, it does not guarantee access. Affordable connectivity, digital literacy, and trust are as critical as bandwidth or device availability. Low-income and rural households face high costs and low speeds. Elders, immigrants, and linguistic minorities must navigate interfaces designed without them in mind. People with disabilities confront platforms that fail to meet basic accessibility standards. The shortcomings of digital infrastructure reflect deeper choices about whose experiences and needs are built into system design. The question is not only technical but also moral: for whom is infrastructure, and who decides?

The gaps in trust and access are further widened by the sheer volume of misinformation circulating across social media and public communication systems. Cinelli et al. (2019) describe how "fragmented information ecosystems" enable the rapid spread of false alerts, contradictory messages, and unverified claims. Communities that face historical underinvestment or surveillance often experience these harms more acutely. For many residents, particularly Black, Brown, Indigenous, immigrant, and rural communities, mistrust of digital systems is not some form of ignorance. It is a rational response to earlier experiences of data misuse, profiling, or policy neglect. Each inaccessible interface, authentication error, or inaccurate alert reinforces the perception that digital tools serve others rather than them.

Rebuilding trust requires more than tightening platform security or improving user interfaces. It requires addressing the underlying power dynamics embedded within digital systems. Participatory governance, where communities have a direct role in shaping how infrastructure is designed, deployed, and monitored, is essential. As the National Academies and numerous digital justice scholars have pointed out, trust emerges from shared decision-making rather than from public messaging campaigns. Communities must be consulted not only about whether infrastructure works, but for whom it works, under what conditions, and with what protections. When residents see their priorities reflected in system design, the legitimacy of digital infrastructure grows.

A justice-centered approach treats digital public infrastructure as a public good. Like roads and water systems, these new environments should be planned and maintained through democratic processes rather than market contracts. This means broadband cooperatives, community-owned cloud storage, and open-source alert systems built to reflect local contexts. It also entails building redundant mesh networks that can operate when centralized systems fail, ensuring that no community's connectivity depends on a single point of failure.

Globally, examples of equitable digital public infrastructure are emerging that demonstrate how community leadership, inclusive governance, and public accountability shape digital systems that meet local needs. In Kenya, community-led broadband cooperatives, such as those supported by the Association for Progressive Communications (APC) and local mesh network builders, have more quickly extended reliable Internet access to rural households than state utilities or private telecommunications firms (APC, 2020; Rey-Moreno, 2018). These non-employers use local governance structures, shared-cost models, and training programs that build technical capacity and maintain community ownership. Evaluations of East African community networks show measurable improvements in economic participation, access to agricultural advisories, and disaster communication during floods and droughts (Ndung'u & Waema, 2019).

In India, large-scale digital public infrastructure has streamlined benefit transfers, digitized payments, and expanded access to public services for millions (Bhatia & Bhabha, 2017; Khera, 2019). The Aadhaar identity system and the broader India Stack ecosystem are examples of providers. At the same time, researchers have documented risks: exclusion caused by authentication failures, privacy concerns, and limited avenues for redress when digital verification determines access to critical social programs (Drèze et al., 2017). India's experience captures the duality of digital public infrastructure at scale. These systems can broaden access when designed with equity safeguards, yet they can reproduce inequality when implemented without transparent governance or accountability (Ramanathan, 2021).

Other examples reinforce these lessons. In Mexico and Colombia, municipal and community broadband partnerships have expanded low-cost access to underserved neighborhoods through hybrid public–community governance structures that treat connectivity as a public good (Gómez, 2020). In Canada and Aotearoa New Zealand, Indigenous nations have built and governed their own broadband and data infrastructures, embedding cultural values and Indigenous data sovereignty frameworks into operational design (Hudson et al., 2020; Kukutai & Taylor, 2016). Taken together, these cases illustrate a shared principle: leapfrogging without equity accelerates inequality. The speed of deployment or scale of adoption means little when systems reproduce the same exclusions they seek to address. Indicators such as bandwidth or user volume may appear to demonstrate progress, but they can obscure gaps in control, privacy, and long-term sustainability. In digital infrastructure, as in climate adaptation, speed is not progress if it leaves the same communities behind.

Designing inclusive digital infrastructure requires embedding equity metrics from the outset. Funding formulas can prioritize historically excluded communities or regions with documented digital deserts.

Procurement systems can require open standards, transparent data practices, and local hiring to build community capacity and ensure continuity. Civic oversight boards can govern digital utilities the way municipalities manage water or energy, treating connectivity as a public asset rather than a private commodity. These approaches make visible what has long remained invisible: the social contracts encoded in code, and the power relationships embedded in the systems that increasingly structure disaster response, economic mobility, and democratic participation.

As digital systems become the backbone of governance, the question "Who builds it, and who it serves?" grows urgent. Infrastructure shapes power because it decides what flows—and what stops. In the digital era, those flows include information, resources, and relief. Communities that lack control over their digital infrastructure risk permanent dependence on those who do. The just future of resilience will belong to those who claim digital infrastructure not as a service, but as a right.

A Way to Design for Digital Justice, Honoring Community Sovereignty in Tech Futures

If technology defines the terrain of modern life, then justice requires knowing who owns that "land" and who draws the maps to navigate it. Digital justice extends beyond questions of access or efficiency. It demands agency, ownership, and governance grounded in community values. In a world where algorithms shape policy and data dictate resource flows, justice cannot be retrofitted into technology; it must be designed into it from the start. The same philosophy informs a growing wave of community-driven technologies. Local cooperatives are building mesh broadband networks that bypass traditional providers. Civic technologists are creating open platforms for disaster communication and participatory budgeting. These projects illustrate what Rossi (2018) describes as engineering trust through co-governance: designing technology not as a service to consume but as a collective problem-solving process. In this model, trust comes from shared authority rather than from a polished interface.

In my own work, I have observed how digital justice emerges when communities can define and govern technology on their own terms. Early in the COVID-19 pandemic, I worked with partners in West Baltimore to map digital exclusion alongside patterns of COVID exposure. Our goal was not only to visualize gaps but to identify how inequities in broadband access affected a community's ability to receive health guidance, participate in remote learning, or organize mutual aid. The resulting maps became tools for organizing, helping residents and advocates make the case for targeted public investments.

That work led to a series of collaborations focused on practical solutions. Churches and neighborhood associations, some of the most trusted institutions in West Baltimore, led efforts to host community Wi-Fi nodes. These spaces served far more than digital needs. They offered cooling centers during heat events, gathering places for distributing supplies, and hubs where older residents could get support navigating online systems. The National Academies of Sciences (2021) notes that "resilient systems emerge from networks of mutual accountability," and these projects demonstrated that principle. Digital infrastructure strengthened existing social relationships rather than displacing them.

Later, some of my students and I worked with community organizations to help draft a Digital Bill of Rights for Baltimore. We approached this as a collaborative policy process, bringing residents with lived experience of digital exclusion into conversation with technologists and public-sector staff. The resulting document emphasized transparency, privacy, accessibility, and accountability, the expectations residents identified as essential for a just digital future. It also linked digital rights to climate resilience and public health, underscoring that access to trustworthy information systems is now a condition of safety.

The process changed how people thought about technology. Residents moved from being data subjects to policy authors. City staff saw the value of embedding community knowledge early, not only to improve services but to build trust in municipal systems. The same philosophy underpins a growing wave of community-driven technologies. Local cooperatives are building broadband mesh networks that bypass traditional Internet providers, while civic technologists are developing open-source platforms for disaster communication and participatory budgeting. This is about designing technology not as a service to be consumed, but as a process of collective problem-solving.

Digital justice also means rethinking who participates in innovation. Too often, communities are invited into technology projects only after key decisions have already been made—after the data architecture is set, the user experience is designed, and the metrics for success have been defined elsewhere. True inclusion begins upstream, when residents shape the questions, assumptions, and priorities that guide a project from inception [National Academies of Sciences, Engineering, and Medicine (2021)].

Several themes are particularly relevant to digital justice. First, communities build the most durable forms of resilience when they have agency in identifying their own risks and setting adaptation priorities. When institutions rely solely on external models or top-down assessments, they risk overlooking local knowledge, place-based vulnerabilities, and the lived realities that determine whether a system will actually function in practice. Second, we cannot over-emphasize the importance of social capital and social connectedness as critical components of adaptive capacity. Communities with strong internal networks are better positioned to translate information into action, especially during disruptions. Locally generated data, whether collected through sensors, surveys, mapping tools, or lived experience, often captures nuances that formal models miss. When institutions incorporate this knowledge into planning processes, the result is not only more accurate but more legitimate. Legitimacy matters because it strengthens public trust in decisions, increases uptake of adaptation measures, and reduces resistance to new systems.

The Undivide Project's work in Iowa illustrates this dynamic. As an expansion of the Ten States Project, residents engaged with GIS tools to refine hazard data and correct omissions that state-level models had overlooked. Their contributions strengthened not only the datasets but the community's sense of ownership over the adaptation process. This aligns with the concept that participatory data practices improve both the technical quality and social legitimacy of resilience initiatives.

Louisiana provides another example. Local networks maintain open-source platforms that record real-time impacts and document how residents organize in response. Such networks represent a form of "distributed resilience," helping communities recover even when formal systems are delayed or overwhelmed. By capturing community action alongside physical damage, these tools preserve a fuller picture of disaster response, one that values social infrastructure as much as physical infrastructure.

Designing for justice also requires confronting the funding and policy structures that sustain inequality. Short-term pilot grants and venture-driven incentives reward novelty over continuity, often leaving communities dependent on external support once the initial project ends. Justice-centered design demands capacity as continuity. This entails resourcing and governance mechanisms that enable communities to maintain and evolve technologies long after the original developers depart. In practice, this might mean shifting grant criteria toward shared-ownership models or establishing long-term funding for digital stewards within local governments and nonprofits. As one "Ten States" participant observed, "We don't need faster tools; we need tools that remember who they're for." The work of digital justice is ongoing and unfinished, but it is already reshaping the boundaries of resilience. Across sectors, from AI ethics to environmental monitoring, communities are proving that technology can be participatory, accountable, and humane. They remind us that progress is not only measured by innovation's speed but also by its reach, how widely it distributes agency and how deeply it embeds trust.

Conclusion: Building the Future we Want, both Technologically and Politically

The task before us is not to abandon technology but to govern it differently. Communities around the world are already engaged in this redesign. They are building data cooperatives, community-run broadband, open disaster platforms, and AI ethics boards grounded in local sovereignty. These initiatives illustrate a new paradigm: technology as stewardship. They translate abstract calls for inclusion into practical architectures of trust. In this model, resilience is not a feature added after the fact, but coded into how systems are created, shared, and sustained.

The same is true concerning justice in digitization. The most secure and adaptable technologies are those that evolve through community participation. When transparency becomes co-governance, and data becomes a shared resource rather than a commodity, the social contract of technology begins to change.

The question then becomes: *What must be done to build inclusive systems from the ground up?* The answer is deceptively simple. Start with the ground itself. Digital inclusion cannot be achieved from boardrooms or cloud servers alone. It must be rooted in neighborhoods, classrooms, and communities that understand their own risks and strengths. It should also include material support, such as training, resources, and tools to enable communities to co-create and govern their own infrastructures.

Designing equitable technology also requires reframing innovation itself. The current culture of "move fast and break things" is ill-suited to the public good. In resilience work, breaking things often means building trust and slowing down. Co-designing, testing, and iterating with affected communities may take longer at the start, but they create technologies that endure. The work ahead is to align digital governance with the moral architectures of community life, ensuring that the same tools shaping the future are grounded in the values that sustain it.

In the next chapter, the focus shifts from principles of justice to real-world encounters with it. In case studies of communities that have experienced slow-onset exclusion, major disasters, and structural challenges, we can understand how resilience and opportunity can emerge from sovereignty. These case studies illustrate how communities facing long histories of divestment are creating durable, sustainable models that offer a glimpse of what is achievable when innovation is aligned with the people it is meant to serve.

References

Acemoglu, D. (2007). Equilibrium bias of technology. *Journal of Political Economy, 115*(6), 988–1032. https://doi.org/10.1086/528606

AI and Inclusion Institute. (2024). AI, equity, and community governance: Principles for inclusive algorithmic design. https://aiandinclusion.org/

Association for Progressive Communications (APC). (2020). Community networks: Communications infrastructure for people by people. APC. https://www.apc.org

Bernard, C. (1987). Machine design bias: The social origins of technical systems. *Technology and Culture, 28*(3), 564–582.

Bhatia, A., & Bhabha, J. (2017). Between exclusion and access: The impact of Aadhaar on benefits delivery in India. *Information Technology & International Development, 13*(2), 1–16.

Borgmann, A. (2012). *Technology and the character of contemporary life: A philosophical inquiry*. University of Chicago Press.

Cinelli, M., Quattrociocchi, W., Galeazzi, A., Valensise, C. M., Brugnoli, E., Schmidt, A. L., Zola, P., Zollo, F., & Scala, A. (2019). The COVID-19 social media infodemic. *Scientific Reports, 10*(1), 16598. https://doi.org/10.1038/s41598-020-73510-5

Committee for Applied Research in Hazard Mitigation. (2021). *Enhancing community resilience through social capital and social connectedness*. The National Academies Press. https://doi.org/10.17226/26123

Committee on Applied Research Topics for Hazard Mitigation and Resilience. (2021). *Motivating local climate adaptation and strengthening resilience*. The National Academies Press. https://doi.org/10.17226/26261

Drèze, J., Khalid, N., Khera, R., & Somanchi, A. (2017). Aadhaar and food security in Jharkhand. *Economic & Political Weekly, 52*(50), 5–9.

Etzioni, A., & Etzioni, O. (2017). AI assisted ethics. *Ethics and Information Technology, 19*(1), 11–27. https://doi.org/10.1007/s10676-017-9412-0

Federal Emergency Management Agency (FEMA). (2020). *Guide to expanding mitigation: Making the connection to equity*. U.S. Department of Homeland Security.

Federal Emergency Management Agency (FEMA). (2021). *Building alliances for equitable resilience*. U.S. Department of Homeland Security.

Gómez, R. (2020). Public libraries, community networks, and digital inclusion in Latin America. *Library Quarterly, 90*(2), 165–184. https://doi.org/10.1086/707722

Hudson, M., Anderson, J., Calderon, V., & Robinson, R. (2020). Te Mana Raraunga and the governance of Māori data. *Journal of Indigenous Wellbeing, 5*(1), 76–84.

Khera, R. (2019). *Dissent on Aadhaar: Big data meets big government*. Orient Blackswan.

Kukutai, T., & Taylor, J. (Eds.). (2016). *Indigenous data sovereignty: Toward an agenda*. ANU Press.

National Academies of Sciences, Engineering, and Medicine. (2021). *Building resilient communities through equitable climate adaptation*. The National Academies Press. https://doi.org/10.17226/26261

Ndung'u, M., & Waema, T. (2019). Community networks in East Africa: A pathway to digital inclusion. *The Journal of Community Informatics, 15*(1), 34–52.

OECD. (2024). The potential impact of artificial intelligence on equity and inclusion in education. Paris, France: OECD Publishing. https://www.oecd.org/en/publications/the-potential-impact-of-artificial-intelligence-on-equity-and-inclusion-in-education_15df715b-en.html

Ramanathan, U. (2021). The architecture of exclusion: Digital ID systems and welfare governance in India. *Law & Policy, 43*(3), 179–201. https://doi.org/10.1111/lapo.12178

Rey-Moreno, C. (2018). Supporting the creation and scalability of affordable access solutions: Understanding community networks in Africa. *IEEE Communications Standards Magazine, 2*(3), 70–76.

Rose, D. (2012). *The moral life of machines: Technology and ethics in the twenty-first century*. MIT Press.

Rossi, L. (2018). Trust and participation in civic technology: The case for co-governance. *Journal of Digital Social Research, 1*(2), 44–61.

Silva, S., & Kenney, M. (2018). Racial bias and the algorithmic governance of inequality. *Information, Communication & Society, 21*(11), 1581–1599. https://doi.org/10.1080/1369118X.2017.1413087

Strate, L. (2012). If it's neutral, it's not technology: The myth of neutrality in media and technology studies. *Explorations in Media Ecology, 11*(1), 1–16. https://doi.org/10.1386/eme.11.1.1_1

Susić, N., et al. (2024). Artificial intelligence and health equity: Risks, gaps, and implications for global populations. *European Journal of Clinical Nutrition, 24*(7), 1175–1189. https://academic.oup.com/eurjcn/article/24/7/1175/8155339

United Nations Framework Convention on Climate Change (UNFCCC). (2024). *AI and climate action: Opportunities, risks, and challenges for developing countries.* https://unfccc.int/news/ai-and-climate-action-opportunities-risks-and-challenges-for-developing-countries

Zhou, Y., & Kapoor, S. (2024). Artificial intelligence and climate resilience: Navigating risk, equity, and governance in the global south. *AI and Ethics, 4*(2), 1–18. https://link.springer.com/article/10.1007/s43681-024-00452-3

Further Reading and Sources Consulted

Carroll, S. R., Garba, I., Figueroa-Rodríguez, O. L., Holbrook, J., Lovett, R., Materechera, S., Parsons, M., Raseroka, K., Rodriguez-Lonebear, D., Rowe, R., Sara, R., Walker, J. D., Anderson, J., & Hudson, M. (2020). The CARE principles for indigenous data governance. *Data Science Journal, 19*(1), 43. https://doi.org/10.5334/dsj-2020-043

Crawford, K. (2021). *Atlas of AI: Power, politics, and the planetary costs of artificial intelligence*. Yale University Press.

Eubanks, V. (2018). *Automating inequality: How high-tech tools profile, police, and punish the poor—New York*. St. Martin's Press.

Hoffmann, A. L. (2019). Where fairness fails: Data, algorithms, and the limits of antidiscrimination discourse. *Information, Communication & Society, 22*(7), 900–915. https://doi.org/10.1080/1369118X.2019.1573912

Leonelli, S. (2019). Data governance by design: How to build FAIR data systems that work for people and the planet. *Nature Genetics, 51*(9), 1306–1309. https://doi.org/10.1038/s41588-019-0494-4

López, S. A., & Pine, K. (2022). Design justice in practice: Centering marginalized voices in civic technology. *ACM Interactions, 29*(4), 30–35. https://doi.org/10.1145/3532357

Organization for Economic Co-operation and Development (OECD). (2019). *OECD principles on artificial intelligence*. OECD Publishing. https://doi.org/10.1787/ea362f40-en

Pasquale, F. (2020). *New laws of robotics: Defending human expertise in the age of AI*. Harvard University Press.

Rainie, S. C., Kukutai, T., Walter, M., Figueroa-Rodríguez, O. L., Walker, J., & Louis, R. P. (2019). Indigenous data sovereignty. *Advances in Information Science, 45*(3), 1–7. https://doi.org/10.1002/asi.24164

Raji, I. D., & Buolamwini, J. (2019). *Actionable auditing: Investigating the impact of publicly naming biased performance results of commercial AI products*. In *Proceedings of the 2019 AAAI/ACM Conference on AI, Ethics, and Society* (pp. 429–435). https://doi.org/10.1145/3306618.3314244.

United Nations Educational, Scientific and Cultural Organization (UNESCO). (2021). Recommendation on the ethics of artificial intelligence. UNESCO. https://unesdoc. unesco.org/ark:/48223/pf0000381137

Whittaker, M., Alper, M., Crawford, K., Dobbe, R., Fried, G., Kaziunas, E., West, S. M., Richardson, R., Schultz, J., & Schwartz, O. (2018). *AI now 2018 report*. AI Now Institute.

Williams, B., & Kindra, G. (2023). Digital justice: A framework for equity-centered technology policy. *Policy & Internet, 15*(2), 246–269. https://doi.org/10.1002/poi3.345

7

Community Sovereignty and Innovation

While national governments and global institutions often frame debates about climate policy and technological governance, the lived experience of the social contract is frequently negotiated at the community level. Local organizations, Indigenous nations, and neighborhoods often confront the consequences of divestment and innovate in response to them, long before these issues appear in national policy discussions.

In this chapter, we have five solutions from five different communities. As we have explored the complexities of the digital divide, climate resilience, and the historical inequities that underpin these challenges, a common thread has emerged: *the impact on communities*. These impacts and stories are also the reason for the intentional use of the term "divestment." Merriam Webster's first definition of the term is "to deprive or dispossess, especially of property, authority, or title," and the second order definition is "to take away from a person."[1] Any of the uses of

[1] https://www.merriam-webster.com/dictionary/divest. Further explanation: The *vest* in *divest* is a close relation of the kind found in closets—its origin is Latin *vestis* meaning "clothing, garment." (*Vest* has the same source and first appeared in English as a verb in the fifteenth century meaning "to put on garments or vestments.") *Divest* today mostly appears in legal and

M. C. Sanders, *Climate and Code*, https://doi.org/10.1007/978-3-032-28032-9_7

the word can apply to communities that were redlined and excluded from access to beneficial and profitable public services and infrastructure. In the case of Indigenous communities, divestment can be used to describe being removed from territories that were the epicenter of cultural and economic life to spaces that lacked investment and were deliberately and repeatedly devalued. Each of these processes may have happened by virtue of law and policy in the United States. Globally, certain communities, such as the one we studied in Indonesia, experience similar impacts from discriminatory actions that span the spectrum from informal to formal government interventions. The result is the same. That lack of access and ability to participate extended to digital infrastructures and the information that comes with them—including critical climate information. These stories of divestment are not the end of the story for divested peoples.

While it is crucial to understand the underlying issues, it is equally important to recognize the solutions and initiatives emerging from communities themselves. So now, we highlight the remarkable efforts of community leaders who are developing their own solutions and driving innovative approaches to some of the problems we have identified throughout this book.

Washington, DC, is the nation's capital and its policy epicenter. It rivals New York for importance on the international stage. Four of the wealthiest counties in the country are located in its suburbs (U.S. News & World Report, n.d.). None of this has made its residents immune to the impacts of digital divestment. Southeast DC, particularly the area known as "Ward 8," was heavily redlined and sited for commercial uses throughout its history. Poor investment in electrical infrastructure in a historically Black neighborhood created, at best, a shaky physical foundation for Internet access. In New Orleans,

---

business contexts about a formal removal or loss of something of value; assets that are divested are sold or given away; someone divested of a right officially loses that right. The world's first late sixteenth-century use, however, was more intimately related to its roots: *divest* was then used to mean "to undress or strip especially of clothing, ornament, or equipment." But broader application of *divest* soon followed. In Shakespeare's *Henry V*, the French King Charles is told via messenger that England's King Henry "wills you, in the name of God Almighty, / That you divest yourself, and lay apart / ... the crown / And all wide-stretched honours that pertain ..." (Merriam Webster)

the story of Broadmoor's divestment hits its peak during the aftermath of Hurricane Katrina. Before, it was one of a few mixed-race, low-income neighborhoods in a city deeply segregated by class and colorism(Preservation Resource Center of New Orleans, 2016). Broadmoor took its modern form during segregation in the broader U.S., which did not make it a hotbed of community investment (Hall, n.d.). It was one of the lowest-lying areas of below-sea level New Orleans. And though it sits on drainage canals, it had the lowest amount of flood and other infrastructure investment in the city. In 1995, ten years before Hurricane Katrina, a series of storms flooded the neighborhood and exposed the vulnerabilities in the community. Yet no actions were taken.

Similar to New Orleans and its neighboring community in DC, Baltimore's story is another that is deeply rooted in race, class, and discriminatory policy. Baltimore City is a coastal city acutely at risk of sea-level rise, flooding, and extreme heat. It also has connectivity so poor across technologies that one must save GPS maps offline due to signal drops in low-lying areas and inadequate infrastructure. West Baltimore is the place where the popular HBO Series "The Wire" was set (*The Wire*, 2002–2008). The communities' struggles with drug trafficking would morph into a legendary reputation that is forever frozen in time by a TV series. The depiction does not make clear that Baltimore is one of the most redlined cities in America. This was shown during the COVID-19 pandemic, in which Black residents died and were left unemployed at significantly higher rates than the local or national average (Social Explorer, 2020). This was evident during our team's walk-through of the neighborhood to local sites for installing Wi-Fi towers: uncovered lead from a neighboring facility, mixed with floodwaters, stained the brownstone steps. Bridges cut through the community, cutting off access to public resources and the opportunities that come with that access. Together, these three urban areas create an image of what "black zones" are in Internet-speak. They also show the cascading impacts of risk across indicators, both online and offline.

The two Indigenous communities with whom we engaged are located in rural contexts at the extremes of the Arctic and Equatorial climates. While not forcibly removed from their territories as has been the case with many Indigenous communities in the Americas, these two groups suffered a different kind of discrimination, which manifested in the form of neglect and being ignored. They were rendered invisible and therefore excluded from critical decisions regarding infrastructure or economic development. As society progresses, inclusion in digitization and climate conversations, while improving, remains less advanced than it could be.

Near the Arctic Circle, connectivity was once a premium reserved for militaries and researchers tracking the loss of polar ice. It was not conceivable that a community would want to have access in general, or that it would use that access to collect climate data and make decisions at the household and group levels. When they were connected and learned how to use the platforms associated with Internet access, the Igloolik decided that preserving culture and language was a priority in the face of climate change. They understood that displacement might be something they would need to confront in a climate-impacted future. Remote learning and creating communities online can be a form of resilience. Their work offers the present and future benefits of holding on to identity while creating support networks. Beyond the physical resilience and economic benefits of confronting digital and climate challenges together, there is an aspect of surfacing elegant solutions that can ease transitions and support new work.

As an example of how governments and communities can overcome historical biases and confront climate change and digital divestment, Panama's approach to sustainable relocation by co-leading with an Indigenous community is a unique and instructive process. Lying just north of the Equator, Panama is at risk of land loss due to sea-level rise and extreme heat. Much of its Indigenous population either lives in the difficult-to-connect mountains or on barrier islands that also lack Internet infrastructure. Many people have no cell phones or ways to connect and receive notifications and information. Despite these circumstances, it was technology, mapping, and crowdsourcing that led to an equitable, sustainable relocation process. The action of sharing leadership with Indigenous groups, connecting them to technology, and then leaving the capacity with the community is a unique lens into how Internet tools and culturally aware behavior can be leveraged to successfully confront climate risk.

The recommendations and critiques in this chapter are meant to inform about solutions within the United States and internationally. They represent community-led solutions. More importantly, they are designed to suggest holistic changes to the traditional "top-down" approaches used in emergency management, disaster science, and resilience communities. For each of the triumphs explored below, there are also lessons to be learned about how not to approach these various themes in current practice. One can compare the rapidity and fluidity of Panama's sustainable relocation of an Indigenous community, where they were co-designers of the plan, with the state of Louisiana's nearly ten-year process of doing the same (WWNO, 2016; Isle de Jean Charles Resettlement, n.d.). The latter did not initially center on the Indigenous communities, nor was it culturally aware. The result was delayed after numerous mediations.

One of the reasons for the Broadmoor case study is to highlight the deliberate decision to destroy a neighborhood without input from communities, businesses, or other stakeholders. In other places, the conflict with larger societal structures need not be as intentional. In most cases, the slow development of programs to help communities address climate change, challenges to equitable funding, or simply the slow but deliberate nature of any activity from the federal government hampers progress.

The most important common thread in all of these solutions is not just the use of technology, the Internet, or resilience building. It is communities and how they interact with different circumstances and collaborators. It is also their ability to solve problems for themselves.

Current Context

Before diving into the case studies, let us begin by examining the current context surrounding community-driven solutions in the realms of digital divide and climate resilience. While the existing body of work has shed light on the challenges faced by marginalized communities, there is a growing recognition of the need to center community voices and empower them in the decision-making processes. Scholars and practitioners alike are exploring innovative approaches and seeking to amplify the successes and lessons learned from community-led initiatives. This chapter contributes to this evolving discourse by presenting case studies that highlight the potential for transformative change when communities are given agency and support.

While progress has been made in addressing the digital divide and enhancing climate resilience, regressive thinking and systemic barriers continue to hinder the advancement of equitable solutions. These ideologies perpetuate existing disparities and reinforce the exclusionary practices that disproportionately affect marginalized communities. It is essential to critically analyze and correct these patterns to pave the way for more inclusive approaches.

When we embrace a future-centered perspective, we can better envision a more equitable and sustainable future. This involves not only addressing the immediate challenges but also identifying long-term strategies that prioritize community empowerment, collaboration, and the dismantling of systemic barriers. By shifting our focus from reactive measures to proactive initiatives, we can create lasting change and build resilience within communities. Policy can be grounded in scientific evidence and then perfected by the lived experiences of those impacted by it. We can see it in each of our case studies.

Case Study #1: Broadmoor New Orleans—From Katrina to Community Resilience

Looking back, most scientists agree that Hurricane Katrina was a climate change-driven superstorm, or at a minimum, its destructive capacity was driven by climate change (Climate Signals, n.d.). As a Louisianan, a New Orleanian in particular, it was clear there was something different about the storm and the wreckage it would leave behind. The devastation caused by Hurricane Katrina in 2005 exposed the vulnerabilities and inequities in New Orleans. The neighborhood in this case study is one whose story I know well, and much of my late childhood and adolescence was spent there. Broadmoor was always a neighborhood "in the middle." Geographically, in the middle of New Orleans and culturally sitting apart from the wealthy, Anglophone enclaves of Uptown New Orleans, the pronounced divestment of the descendants of the enslaved in the Lower Ninth Ward is quite unlike the French Creole neighborhoods responsible for the city's fame.

This positionality makes Broadmoor a microcosm of the broader city and country. That makes it a good case study for the inequities Katrina exposed and the battle to overcome those challenges and put meaning to the words "build back better." In the aftermath of the disaster, community leaders and residents came together to rebuild and reimagine their community. We delve into their journey, exploring how people there transformed seemingly insurmountable challenges into opportunities for resilience. Through a combination of grassroots efforts, innovative approaches to healthcare, and the establishment of a regenerative urban farm, Broadmoor exemplifies the power of community-driven solutions. We examine the lessons learned from their experiences and highlight the potential for collaboration between the emergency management and hazard mitigation communities to learn from and support communities like Broadmoor.

From Green Dotting to Rebirth and Resilience

On January 11, 2006, residents of New Orleans' Broadmoor neighborhood, still grappling with the aftermath of Hurricane Katrina, were confronted with a proposal from the city to transform their community into a park. The Urban Planning Committee of a mayoral commission had recommended giving neighborhoods like Broadmoor four months to prove their "viability" and worthiness of rebuilding. Adding to their shock, a map published by The Times-Picayune showed a green dot covering Broadmoor, signifying that the historically low-lying area was suitable for "greenspace" but not for living. Feeling betrayed by their city government, Broadmoor residents embarked on a journey to save their neighborhood from demolition. They formed the Broadmoor Improvement Association (BIA) and kicked off an advocacy initiative that would become the "Broadmoor Lives" campaign.

Broadmoor, like most of New Orleans, is situated in a bowl-shaped area below sea level. During Katrina, the area was hit with slow but relentless flooding that rendered homes uninhabitable due to electrical system damage and black mold infestations. To put a finer point on the description of it being a microcosm of New Orleans as a whole, its demographic composition is 68 percent Black, 26 percent White, 4 percent Hispanic, and 1 percent multiracial residents, according to the 2000 U.S. Census (U.S. Green Building Council, 2014). The average household income was $50,000.

What set this group of neighbors apart is that rather than succumbing to despair, they rallied around the green dot symbol. They transformed it from a sign of potential destruction to a logo representing their determination to rebuild. To counter the city's proposal and regain control of their community's future, Broadmoor residents organized a 300-person rally and flooded the neighborhood with green banners and signs proclaiming "Broadmoor Lives." This initial activism marked the beginning of an era of unprecedented community engagement that would last for more than a decade. In 2008, residents celebrated the progress made by cutting a cake in the shape of a green dot, symbolizing their resilience. Ten years later, they celebrated the opening of a community center, then a fitness trail, and more improvements as the post-Katrina years passed. Almost twenty years later, Broadmoor stands as a symbol of community resilience.

Forming Critical Partnerships

The BIA played a crucial role in mobilizing residents and organizing the planning process for the scattered community. Although many core members had no prior experience in redevelopment, they took the lead in creating a recovery plan. It drew the attention of national academics and think tanks. Recognizing the importance of local leadership, Harvard University's Kennedy School of Government and the Clinton Global Initiative (CGI) partnered with the BIA to co-create a redevelopment plan, a necessary component to prove its "viability." Residents divided the neighborhood challenges into three subsections and sets of issues, then they made sure to have diverse representation in the decision-making process (Harvard Kennedy School Case Program, 2008).

Central to Broadmoor's 320-page redevelopment plan was the establishment of an "education corridor," featuring a rebuilt community library and an elementary school. The efforts of Broadmoor's residents and their partnership with CGI led to Andrew H. Wilson Elementary being selected for fast-tracked renovation in the city in 2007. It transformed and provided a renewed essential service to Broadmoor. By 2010, the school moved into a newly renovated LEED Gold building featuring solar panels, rain gardens, sustainable design elements, and computer labs. Two years later, the LEED-certified Keller Library and Broadmoor Community Center opened, providing educational programs, community gathering space, and wellness activities (Broadmoor Improvement Association, n.d.). Two examples of the connection and support offered at the community center are yoga and wellness classes, and a clothing store that provides consultations and other support to residents returning to work.

The BIA did not overlook additional traditional elements of resilience related to disasters, food, and housing security. Broadmoor now has a community garden and pantry that serves food-insecure residents daily, but is capable of supporting the community in the critical "first 72" hours after a disaster. There is also a tool and supply lending library. Residents can check out tools and supplies from the library to perform house repairs and upgrades to mitigate hazards and maintain the safety and security of their homes.

Leveraging New Tools to Build Resilience

The citizens of Broadmoor were also forward-thinking in not just considering recovery as traditionally imagined, but also how to bring other amenities to the community to make that recovery sustainable. One thought was the importance of robust Internet infrastructure and community Wi-Fi in fostering resilience and enabling essential services. As part of its comprehensive recovery plan, Broadmoor recognized the significance of reliable Internet access for its residents and championed improved connectivity.

Community leaders understood and still understand that in the digital age, Internet access is vital for accessing information, resources, and services, especially in times of crisis. Recognizing the potential of telehealth services to support the mental health recovery of its community, the BIA prioritized the development of better Internet infrastructure. This enabled residents to access healthcare professionals remotely and on a sliding scale. This is particularly beneficial for those who face challenges in accessing healthcare facilities due to geographic, financial, or mobility constraints.

Community Wi-Fi ensured that Broadmoor residents could stay connected and informed. The BIA uses online resources to share crucial information about its activities, resources, and events with confidence, knowing that residents would be able to receive it. This facilitated efficient communication, enhanced community engagement, and promoted transparency in the recovery process. It also ensured that all residents, regardless of their economic background, had equal access to digital resources and opportunities. This inclusivity helped narrow the information gap and ensured that vulnerable community members were not left behind in the recovery process. The benefits of making information access democratic and open range from having open lines to city hall, emergency management, and other public services for situational awareness and civic engagement to being able to access public and individual assistance resources after an event. The connection allowed the community to create in-community work. The community center employs social workers, administrative professionals, facilities and inventory managers, and fitness instructors. More people are either employed or are volunteers in the garden and tool lending library. Financial flows and emotional and intellectual capacity stay within the community, reinforcing the all-important social factors of resilience (Haldane et al., 2023).

The motto "Better Than Before" captured the community's commitment to surpassing pre-Katrina conditions. Broadmoor's determination to turn ideas into action led the community to leverage the power of the Clinton Global Initiative (CGI), a platform focused on solving global challenges through cross-sector partnerships (Clinton Foundation,

2015). Broadmoor's diverse group of organizations made a bold commitment through CGI to empower local leadership and secure resources to support the redevelopment plan.

Broadmoor's success can be measured in many ways, but is exemplified by its impressive repopulation rate, estimated at nearly 90 percent. Although the neighborhood's population in the 2010 census remained lower than its pre-Katrina numbers, efforts to attract residents have continued. This showcases the power of community-driven initiatives and cross-sector partnerships in building resilience. Broadmoor's advocacy for better Internet infrastructure and community Wi-Fi showcases the forward-thinking approach necessary in building resilient communities. By prioritizing digital connectivity, policymakers and stakeholders can ensure that communities have the tools and resources they need to navigate challenges effectively. The "Broadmoor Lives" campaign, which galvanized residents, and the commitment to local leadership through partnerships were crucial elements of the neighborhood's successful recovery. The creation of the BIA brought sustainability and a portfolio of new public services into the community. It demonstrated how post-disaster opportunities can be leveraged effectively and equitably.

The collective efforts of residents, the Broadmoor Improvement Association, Harvard Business School, and Clinton Global Initiative transformed into a comprehensive, future-focused plan that did not just save a neighborhood but transformed it into a thriving and empowered community. Broadmoor's story serves as an inspiring example for other communities, highlighting the significance of grassroots efforts, inclusive planning, and partnerships in fostering resilience and long-term recovery.

Lessons from Broadmoor About Community-Led Disaster Recovery and Hazard Mitigation

Broadmoor's journey to resilience offers valuable lessons for the emergency management and hazard mitigation communities, as well as policymakers. By examining Broadmoor's example, these stakeholders can gain insights into effective strategies for community engagement, grassroots empowerment, and long-term recovery.

Emergency management professionals should recognize the importance of community engagement and involvement in all stages of disaster preparedness and recovery. The community redevelopment plan's success was largely attributed to the active participation of its residents, who took charge of the process and advocated for their neighborhood's future. This highlights the need for emergency management agencies to foster partnerships and collaborative decision-making with local communities, ensuring that their voices are heard and their expertise is acknowledged.

Furthermore, hazard-mitigation communities can learn from Broadmoor's emphasis on long-term planning and proactive measures. The establishment of an education corridor and the commitment to sustainable infrastructure (physical and digital) demonstrated Broadmoor's dedication to building a better and safer community.

Policymakers should understand the importance of incorporating hazard mitigation strategies into recovery plans from the outset, rather than solely focusing on immediate relief efforts. By prioritizing resilience and sustainability in rebuilding projects, policymakers can mitigate future risks and create more resilient communities.

Another crucial lesson from Broadmoor is the significance of cross-sector partnerships. Broadmoor's collaboration with the Clinton Global Initiative and the engagement of Harvard University's Kennedy School of Government showcased the power of partnerships in leveraging resources, expertise, and support. Policymakers should recognize the value of forging alliances with academic institutions, nonprofit organizations, and the private sector to maximize the impact of recovery and mitigation efforts. By fostering collaboration and leveraging diverse resources, policymakers can develop comprehensive, sustainable solutions to address the complex challenges posed by disasters.

Overall, Broadmoor's example highlights the importance of community empowerment, long-term planning, and cross-sector partnerships in the field of emergency management and hazard mitigation. By integrating these lessons into their practices, professionals and policymakers can enhance their capacity to respond effectively to disasters, promote community resilience, and build sustainable communities capable of withstanding future hazards.

Case Study #2: Anacostia, Washington, DC, Building Economic Opportunity, Justice, and Resilience: The Ward 8 Digital Collective as a Game Changer for Urban Communities

In the Anacostia neighborhood of Washington, DC, community leaders have been at the forefront of addressing the digital divide and fostering digital equity. We examine ongoing efforts and initiatives to expand access to technology resources, digital literacy training, and economic opportunities. However, we also acknowledge the challenges faced by the community, including limited resources and systemic barriers. By understanding the lessons learned and ongoing challenges, we can identify pathways forward and advocate for policies that support equitable digital access and empowerment.

Ward 8, located in Washington, D.C., has long grappled with systemic challenges that have hindered economic opportunity and perpetuated inequalities. Historical issues have deeply affected the community, leading to concentrated poverty, limited access to quality education and healthcare, high unemployment rates, and a lack of economic development. Gentrification is an ongoing challenge, with Ward 8 residents watching as new high-rises and higher housing costs push their peers out of Washington, DC. Many feared the inevitable would come to their communities. However, in recent years, Ward 8 has emerged as a center of community-driven resilience, with initiatives such as the Ward 8 Community Economic Development (W8CED) project aimed at addressing these long-standing challenges. At the heart of this effort is the Ward 8 Digital Collective, an innovative online platform that has the potential to revolutionize urban resilience and anti-gentrification efforts.

Historical Context: Redlining and Its Impact on Ward 8

To understand the challenges faced by Ward 8, it is crucial to recognize the historical context. During the era of redlining in the mid-twentieth century, Ward 8, with its predominantly African-American population, became one of the areas most affected. The legacy of disinvestment and neglect created significant barriers to economic mobility and contributed to the systemic challenges the community faces.

In response to these historical injustices and the ongoing struggles faced by Ward 8 residents, the W8CED initiative was launched. This community-led effort sought to develop a comprehensive economic development plan grounded in the community's needs and aspirations. After years of engagement and dialogue, the initiative produced "A Dream Deserved: Realizing Our Collective Emergence," a report that outlined the community's priorities and aspirations (Ward 8 Community Economic Development, n.d.).

Those aspirations represented a desire to not simply fight against erasure or gentrification, but to build an inclusive, connected community. The plan, similar to that in Broadmoor, was meant to convince a reticent city hall that alternative thinking about neighborhoods and their "viability" can be seriously flawed. In "A Dream Deserved," the community outlines a plan to reinvest in and support small businesses. This would not only spur business ownership, bringing tax revenue and monetary flows back into the area. Another area of concern is the reinvestment of a different kind, bringing parental and community involvement back into the local elementary, middle, and high schools. The people of Ward 8 sought to reimagine a pipeline that once physically cut young people and businesses off from success, creating one that redirects energy back into the neighborhood and youth. Energizing the community around young people is a way to reduce negative impacts, improve school performance, and mitigate outbound migration, or "brain drain," in communities. All of these cut against resilience by any definition.

A marquee aspect of the entire W8CED initiative is the Ward 8 Digital Collective. This online platform has the potential to be a game-changer for urban resilience and anti-gentrification efforts. It provides streamlined access to essential community services, such as community meetings, small-business loans, and other information, through a cloud-based platform. It has a membership component that allows community members to log on and participate. By providing a space for community members to connect, share information, and collaborate, the collective aims to empower residents, strengthen community ties, and amplify the community's voice in decision-making processes. These are some of the priorities and solutions put forth in the collective (W8CED, 2023).

1. **Economic Opportunity and Justice.**
   Through the platform, residents can access information on job opportunities, business resources, and support services. It provides a space for local businesses, nonprofits, and institutions to engage with residents, fostering economic development and creating pathways to prosperity within the community.
2. **Community Empowerment and Resilience.**
   The digital platform catalyzes community empowerment and resilience, with a focus on seniors who have time to be civically engaged for the community, business owners, youth, and returning citizens. By facilitating communication and coordination among residents, the collective enables collective problem-solving and amplifies community-led initiatives. This promotes a sense of ownership and agency, allowing residents to shape the future of their community and build resilience in the face of challenges.
3. **Anti-Gentrification Efforts.**
   The Ward 8 Digital Collective also plays a critical role in anti-gentrification efforts. Residents have the opportunity to learn about zoning, siting of facilities, and other land use decisions in a way that is timely and accessible. Language is broken down in ways that everyone is able to understand what is being proposed. By providing a platform for residents to organize, advocate, and hold stakeholders accountable, the collective helps safeguard the community's interests and cultural identity. It supports efforts to mitigate displacement, protect affordable housing, and ensure that economic development benefits existing residents rather than contributing to their marginalization.

Game Changer for Urban Resilience and Anti-Gentrification Efforts

The Ward 8 Digital Collective has the potential to revolutionize approaches to urban resilience and anti-gentrification efforts. By leveraging digital tools and community-driven platforms, it empowers residents to participate in decision-making processes actively, address systemic challenges, and shape their community's future. The collective's focus on economic opportunity, justice, and community empowerment aligns with broader movements seeking to create more equitable and resilient cities.

The success of the Ward 8 Digital Collective could serve as a model for other urban communities facing similar challenges. By harnessing technology, engaging residents, and fostering collaboration, similar initiatives could be replicated in other neighborhoods grappling with gentrification, disinvestment, and systemic inequalities. The digital collective concept has the potential to transform the dynamics of power, strengthen community bonds, and ensure that marginalized voices are heard and included in urban planning and development processes.

The Ward 8 Digital Collective represents a transformative approach to building economic opportunity, justice, and resilience in urban communities. By addressing the challenges faced by Ward 8 as a community and for the community, the W8CED initiative recognizes the need for community-driven solutions. The digital collective provides a space for residents to connect, collaborate, and advocate for their needs, fostering economic development, community empowerment, and anti-gentrification efforts. As a potential game-changer, this innovative approach could inspire and inform similar initiatives across cities, paving the way for a more equitable and resilient urban future.

Challenges Facing the Anacostia and the Ward 8 Community

Digital Divide and Access to Technology: One of the significant challenges faced by the Ward 8 community is the digital divide, which refers to the disparity in access to technology and digital resources. Many residents lack reliable Internet access, digital literacy skills, and access to devices necessary to participate in the Ward 8 Digital Collective fully. Bridging the digital divide requires concerted efforts from the city of Washington, D.C., to allocate resources for new digital infrastructure, expand broadband access, and provide training programs to enhance digital literacy skills among residents. This challenge highlights the need for equitable access to technology and digital resources to ensure that all community members can benefit from the opportunities presented by the Ward 8 Digital Collective.

Vulnerability to Climate Change: The Ward 8 community's location on the Anacostia watershed exposes it to climate change-related challenges, such as riverine flooding and other environmental impacts. This is another space where climate change and the digital divide intersect to increase a community's overall risk profile—even one with groundbreaking work already happening. A strong Wi-Fi signal is required to support the air and water quality monitoring needed to understand what is happening with the river in terms of sea-level rise and contamination. While the community is eligible to receive federal support funding for broadband, the amount of infrastructure that needs to be planned, built, and deployed in neighborhoods is a years-long process (Benton Institute, n.d.). As climate change intensifies, extreme weather events, sea-level rise, and increased precipitation pose threats to the community's infrastructure, housing, and overall well-being. In the meantime, measuring the exact impact of these hazards will be difficult without an effective middle or last-mile Internet solution in this community. There are projects where the Ward 8 collaborative is partnering with outside organizations on a variety of climate mitigation and environmental justice programs. In addition to Internet infrastructure that allows for more creativity, like the digital collaborative, combining efforts among the community, city authorities, and environmental organizations will be crucial to building the community's resilience to climate change.

Displacement and Gentrification Pressures: As the Ward 8 community undergoes revitalization and economic development efforts, there is a concern about potential displacement and gentrification pressures. These issues can arise when new investments and development projects lead to rising property values, increased costs of living, and the displacement of long-time residents. To address this challenge, the Ward 8 Digital Collective and other community-led initiatives need to prioritize anti-displacement measures, affordable housing strategies, and policies that ensure the community's cultural identity and socioeconomic diversity are preserved. Collaborative engagement between community organizations, developers, and policymakers is necessary to strike a balance between economic development and social equity.

Equity in Resource Allocation: Ensuring equitable resource allocation is a fundamental challenge that the Ward 8 community must confront. Historically marginalized communities like Ward 8 have often experienced disparities in access to resources and public services. The city of Washington, D.C. must prioritize investment in Ward 8, allocating resources to address the community's specific needs in areas such as education, healthcare, infrastructure, and economic development. This challenge calls for transparent and inclusive decision-making processes that prioritize the voices and priorities of Ward 8 residents, enabling them to actively participate in resource allocation discussions.

While the Ward 8 Digital Collective holds great promise for the community's economic opportunity, justice, and resilience, several challenges need to be addressed to ensure its success. Overcoming the digital divide, strengthening climate resilience, mitigating displacement and gentrification pressures, and promoting equitable resource allocation are critical areas that require collaborative efforts from the community, city authorities, and various stakeholders. By actively addressing these challenges, the Ward 8 community can build a more inclusive, resilient, and prosperous future that benefits all of its residents.

Lessons we Can Learn from Ward 8 of Washington, DC

One of the lessons we can all learn from Ward 8 is the importance of recognizing community knowledge and agency. Particularly when it comes to mostly BIPOC or historically marginalized groups, there is a tendency to consider them beneficiaries of subject matter expertise rather than holders of that expertise. The development of a comprehensive plan and the creation of the digital collective demonstrate agency in the form of being able to identify challenges and organize around them. There is also leveraging community-based assets and expertise to see the digital collective created and adjusted for the community's needs. Reframing how we view certain communities is an important step to being able to co-create and generate solutions with them.

Case Study #3 Baltimore, MD Case Study: Bridging the Digital Divide through Community Wi-Fi Networks

Sandtown, Cherry Hill, and Forest Park are all known for being part of one of the toughest and most impoverished areas in Baltimore. Baltimore as a whole is a former industrial capital known as much for its need for reinvestment as for its name, "Charm City." It is also known for divisions between law enforcement and the community, poverty, and a lack of access to public goods, the Internet being one of them. It also has a large population of people under the age of 25. All of these factors come together to create a unique approach to connecting and building resilience. The process started with understanding that West Baltimore is an "urban black zone," a term for an area that lacks quality access to the Internet or has no access but is in an urban area that is otherwise highly connected and digitized.

In February 2020, a group of high school students in West Baltimore embarked on a training program to configure networking equipment, little knowing that the world was on the brink of a deepening COVID-19 pandemic. These students were about to play a crucial role in bridging the digital divide in their community. This case study explores their journey and the transformative impact of community networks in providing reliable and affordable connectivity to underserved neighborhoods.

This was also a journey that would begin the work that led to *The Undivide Project's* formation and part of the inspiration for this book. The project in West Baltimore started as one to create community Wi-Fi and study how they could reverse the impacts of the digital divide. As with many research and philanthropic projects that begin from the outside, the presence of the teams was met with resistance. The onset of the pandemic paradoxically helped the community network initiative gain traction. The urgent need for Internet access became evident as the world shifted online and COVID-19 divided us into groups of essential workers and those with online access. During this time, we came to learn how necessary it is for education, work, and healthcare to be available online. They are critical public goods, and as society digitizes, they have to be widely available in all spaces that encompass the whole of society. COVID-19 would hit Baltimore and West Baltimore particularly hard. One of the reasons was the overrepresentation of "essential workers" living in the neighborhood. Then there is the presence of a lead facility in the neighborhood, contributing to respiratory issues and other public health issues in the communities. Because Baltimore is also at high risk for sea-level rise and flooding, this same lead is mixed with the water to create environmental justice and climate risk. To name and seek a remedy for any of these issues, measurements are a requirement. The community needed sensors, and those sensors needed a Wi-Fi connection.

It became clear that these young people were being called upon to tackle multiple challenges at once. The first training began in February of 2020. Because masking and working outside were relatively safe at that point in the pandemic, they implemented that learning fairly quickly. The students installed the first community Wi-Fi antenna at the New Song Academy (cite or footnote firsthand account), a community school in Baltimore. This installation provided a much-needed lifeline for families in the area.

The free Wi-Fi signal enabled parents to complete essential tasks such as forwarding documents, checking emails, and scheduling medical appointments. Students in the community were able to attend class at school, but online and safely, like kids in other neighborhoods. Workers at the food bank, also located at the same school, were able to coordinate orders and safe food pickup in the community again. The first impact of bridging the digital divide and using the power of the Internet to push past one of the legacies of intentional exclusion was evident within the first month of work.

Expanding the Network and Empowering the Community

To expand the reach of the community network, the group secured a grant from the Internet Society's Community Networks project. This funding allowed them to order additional equipment and set up more Wi-Fi hotspots. The installation of hotspots continued at various locations, including Elev8 headquarters and Patapsco Elementary School. However, it soon became apparent that the trained high school students alone could not provide the necessary workforce. To address this, the initiative partnered with a local trade school and employment agency, NPower, to recruit part-time workers. By September 2020, four young NPower graduates were hired as full-time consultants and employees to build out the network further. These three installations covered almost 100 families and several businesses.

The community network initiative aimed not only to provide Internet access but also to empower community members. By incorporating a workforce development plan, the project sought to train individuals and establish a local worker-owned cooperative. This approach aimed to create a sense of community ownership and foster long-term sustainability. Elev8, the United Way of Central Maryland, RowdyOrb.it, and the Internet Society played vital roles in imparting both soft and hard skills necessary for community network installation. Now this project has addressed another component of digital divestment, access to employment and opportunities for community ownership.

The impact of the community network initiative on individual lives was substantial. Participants who previously lacked reliable Internet access and clear career pathways were able to transition into sustained technical and leadership roles within the network. Access to Wi-Fi expanded educational and professional opportunities, enabling individuals to develop both technical competencies and broader professional skills. Community stewards emphasized that connectivity helped level access to opportunity, allowing them to contribute meaningfully to local economic activity and community well-being.

Finding New Ways to Expand Coverage and New Community Connections

Recognizing the vital role played by churches in supporting underserved communities, the initiative extended its efforts to provide Wi-Fi hotspots at the AME and United Methodist Church in Cherry Hill. The involvement of the United Way of Central Maryland further amplified the project's impact through additional funding, long-term backhaul solutions, and ongoing compensation for those involved in the network installation.

Building upon their success in community centers, the initiative began expanding the network to homes. The goal was to connect one home per day, reaching a total of 210 homes in the first year. Each hotspot covered a half-mile radius, ensuring maximum coverage for selected households. Access and equipment remained free for participating households and users during the first year. The long-term vision was to develop a sustainable cooperative business model that would be owned and managed by community members while keeping costs as affordable as possible. The community-led approach and unique incentives fostered reliable and affordable connectivity, addressing the gaps left by the existing market.

Now that coverage was in place and deep community ownership and cohesion were the centerpiece of the installation and maintenance elements of the project, the opportunity to deal with the environmental and climate justice issues had come. This element of the work was also led by young people. Students at Johns Hopkins School of Public Health began working to understand where and how sensors could be installed with the Wi-Fi network. Another group of students from Georgetown worked with the New Song Academy and other students to identify and map areas of potential contamination. The group also did an asset mapping exercise to understand who would be able to take long-term responsibility for parts of the project. Following the same model as the installation process, co-design happened with the community. The same is true for co-learning about how to place sensors and collect and interpret data. These combined efforts would result in a series of data-driven advocacy projects. This collaboration is still happening today. The Baltimore community network initiative stands as a powerful example of how grassroots efforts can bridge the digital divide and empower underserved communities to address multiple challenges. It also shows how climate and environmental risk intersect with public health, economic stability, and digital divestment. The latter is the starting point for true risk reduction.

Perspectives and What we Can Learn about Possibilities for Other Communities

In addition to bridging the digital divide, community networks like the one in Baltimore contribute to building community resilience. The COVID-19 pandemic highlighted the importance of resilient communities that can adapt and respond to crises. During the pandemic, the Baltimore community network proved to be a valuable asset for the community. It facilitated remote learning for students, enabled access to telehealth services, supported in-community jobs and remote work, and connected residents to vital resources.

Community networks can play a vital role in disaster and climate change mitigation. The network was not only key to measuring the health and safety risks associated with the lead facility, but it also helped measure and predict flood risk. This information was and continues to be used to direct and resource mitigation projects. Community networks, being locally owned and operated, can continue to provide connectivity, enabling communication, coordination, and access to emergency services. When there is an event, a community that would have typically been cut off from digitally based alerts and early warning systems can now receive life-saving information. Then there is access to hazard mitigation funding and, in the event of a disaster, being able to apply for critical funding.

In times of crisis or emergencies, community networks can serve as platforms for mutual support and resource sharing. They enable neighbors to connect, share information, and provide assistance to those in need. The network becomes a tool for community organizing, enabling residents to mobilize and address common challenges collectively. Moreover, community networks contribute to social and economic resilience. They provide opportunities for individuals to acquire digital skills, participate in the digital economy, and access online educational resources. This empowerment leads to increased economic opportunities and improved quality of life for community members.

Furthermore, community networks strengthen social cohesion and civic engagement. They create spaces for collaboration, information sharing, and community-building. By involving residents in the planning, implementation, and management of the network, community networks foster a sense of ownership, pride, and collective responsibility. This engagement builds social capital and strengthens the bonds within the community, enhancing its overall resilience. It also gave community members a sense of agency and access to public resources to petition for more resilience-building tools.

Community networks not only bridge the digital divide but also contribute to community resilience. By providing reliable connectivity, facilitating access to critical services, fostering social cohesion, and enabling local economic development, community networks empower communities to be more resilient in the face of crises. As communities continue to face various challenges, investing in community networks becomes an essential component of building resilient and thriving communities.

While these urban examples reveal how connectivity can address deeply entrenched structural inequities, rural and Indigenous communities face a different configuration of challenges. Here, distance from decision-making centers, sparse infrastructure, and climate extremes shape how the digital divide intersects with resilience. The next two case studies, Igloolik in Arctic Canada and the Guna community in Panama, show how technology can be harnessed in these contexts not only to address immediate risks but to preserve culture and co-create climate adaptation pathways.

Case Study #4 Igloolik, Canada—Leveraging ICT for Resilience, Opportunity, and Climate Change Response: A Case Study of Igloolik, Canada

The economic resilience and contributions to the social fabric of communities that come with Internet access are exemplified in Igloolik. Igloolik, the primary hamlet on Igloolik Island, is situated within the Canadian territory of Nunavut. The Indigenous community of Igloolik, known as Iglulingmiut, takes its name from the important settlement site Iglulik, which means "it has houses." (The Canadian Encyclopedia, 2015). The Iglulingmiut's country stretches from Igluligaarjuk on northwestern Hudson Bay, northward along Melville Peninsula, and across the northern third of Baffin Island. This geographic context is essential for understanding the unique challenges and opportunities faced by the community. While not facing having to cede territory like tribes in the United States, the Iglulingmiut have faced intentional and unintentional divestment. Because of the remote location of their community, access to public services and civic engagement has been limited. The terrain is harsh and unforgiving to many kinds of infrastructure, particularly Internet infrastructure. Harsh situations often breed creativity. In recent years, Igloolik has creatively embraced information and communication technologies (ICTs) to build resilience, create economic opportunities, and address the impacts of climate change.

Enhancing Connectivity and Economic Development

Despite its remote location, Igloolik has recognized the transformative power of ICTs and has actively embraced connectivity (Young, 2019). In 2009–2010, the Government of Nunavut estimated that 59% of households across Nunavut, and 49% of households in Igloolik, had Internet access. Since then, the availability and accessibility of the Internet have increased, with most Iglulingmiut having Internet access at home and/or work. This has opened up new avenues for economic development and educational opportunities.

Internet access has become an essential tool for various aspects of daily life in Igloolik. For instance, the Internet has facilitated job searches, enabling residents to explore employment opportunities within and beyond the community. Job hunters can now perform searches from the comfort of their homes, eliminating the need for the traditional approach of going from building to building to inquire about job vacancies. Digital platforms have also played a significant role in supporting local businesses. Elijah, the owner of a local hotel, emphasized how the Internet has made running his business easier. He utilizes online platforms for advertising, banking, and bill payments, highlighting the efficiency and convenience offered by digital technologies.

Moreover, the integration of new technologies into everyday life has not only bolstered the local economy but also empowered individuals and provided them with opportunities for personal and professional growth. Students can use Internet access to pursue education and future career prospects. The availability of online resources has expanded educational opportunities, enabling residents to take classes, access learning materials, and gain the knowledge and skills necessary for career advancement.

Improving Social Cohesion and Cultural Preservation

Igloolik residents have harnessed the power of digital technologies to strengthen social resilience and preserve their cultural heritage. While the Internet has facilitated connectivity with global markets, it has also been used to reinforce local Inuit culture and traditions. Platforms like the "Nunavut Hunting Stories of the Day" Facebook groups have emerged, serving as virtual spaces for Iglulingmiut to share hunting experiences and support traditional Inuit practices (Young, 2021). These platforms foster community engagement, strengthen cultural bonds, and provide opportunities for intergenerational knowledge transfer.

Kingulliit Productions, Inc., an organization in the hamlet, has gained global recognition as an innovator in Indigenous news media. Their work showcases the use of digital technologies to preserve and promote Inuit culture. Through film and multimedia productions, Kingulliit Productions, Inc. documents traditional practices, shares stories, and raises awareness about Inuit perspectives and ways of life. Their initiatives not only contribute to cultural preservation but also create economic opportunities within the community, providing employment for local talent and encouraging the development of creative industries.

Addressing Climate Change Challenges

Igloolik, like many Arctic communities, is directly affected by the impacts of climate change. Rising temperatures, sea-level rise, and changing weather patterns pose significant challenges to the community's traditional livelihoods, infrastructure, and overall well-being. In response to these challenges, the community got connected and then used ICTs as a means to adapt and respond to the changing environment.

The first step was understanding and sharing information about their changing environment. Using platforms like Facebook and messaging apps, the Iglulingmiut were able to reach out across their geographic separations and engage in conversations about climate change, share knowledge and experiences, and advocate for sustainable practices. From these conversations, ideas about adjusting hunting practices as a mitigation technique arose. The community rethought hunting seasons and the scale of those endeavors. With the introduction of Internet access came the opportunity to create new economic flows. Artisans could set up digital storefronts, and other goods and services unique to their community could be sold outside as e-commerce. So in the face of losing some economic sustenance to climate change, Internet access created new opportunities and resilience practices (Ford et al., 2006).

They also led the Iglulingmiut to address another concern, language loss. With more young people leaving the community for work and education and more elders passing away, the Iglulingmiut faced a loss of language and culture, which is a concern in many Indigenous communities. Digital platforms can play a crucial role in educating and inspiring Inuit youth to navigate the contemporary world while drawing upon their cultural heritage. The community proposed and then co-designed with data architects a platform that could teach Inuit youth their language and cultural heritage. It also includes lessons about how their people have historically survived in the harsh conditions of the Arctic. Some of this traditional knowledge has been shared with scientists. Scientists have also learned to embrace Indigenous knowledge, employing some of the community to help with ice monitoring in the Arctic using a combination of traditional science and Indigenous knowledge and interpretation (Ford et al., 2020).

By showcasing traditional knowledge and practices, the platform not only inspires resilience and adaptation in the face of climate change among that community but also serves as a constant source of knowledge and understanding for other communities.

Challenges Facing Igloolik: Mitigating Digital Divides and Infrastructure Upgrades

While Igloolik has made significant strides in embracing ICTs, the community still faces challenges related to infrastructure and connectivity. The current Internet infrastructure, based on outdated dial-up connections and satellite support, limits the full potential of digital technologies. Igloolik requires infrastructure upgrades to support faster and more reliable Internet access. This will further reduce costs and expand opportunities for economic development, education, and cultural preservation.

A comprehensive survey is necessary to identify the most appropriate technological solutions to improve connectivity in Igloolik. Broadband, more use of satellite Internet, or community Wi-Fi are potential options that could bridge the digital divide and unlock the community's full potential. By addressing infrastructure challenges and ensuring equitable access to digital technologies, Igloolik can continue to thrive, fostering economic growth, educational opportunities, and cultural resilience.

The case of Igloolik demonstrates how an Indigenous community can harness Internet infrastructure and associated ICTs to build resilience, create economic opportunities, and address the impacts of climate change. Through enhanced connectivity, Igloolik residents have leveraged digital technologies to improve economic prospects, preserve cultural traditions, and engage in discussions and scientific research on climate resilience. However, the community still faces infrastructure and connectivity challenges that must be addressed to enable further progress. By prioritizing infrastructure upgrades and bridging digital divides, Igloolik can continue to leverage ICTs to build a sustainable future, empowering its residents to navigate contemporary challenges while preserving its rich cultural heritage.

Lessons we Can Take to Other Rural Contexts

The case study of Igloolik provides valuable insights that can inform approaches to rural resilience building in other communities facing similar challenges. Firstly, it highlights the importance of connectivity and access to information and communication technologies (ICTs) in rural areas. By embracing digital technologies and improving Internet infrastructure, rural communities can unlock economic opportunities, enhance education and skill development, and strengthen social resilience. The Igloolik case demonstrates that even in remote locations, ICTs can bridge the gap between rural and urban areas, connecting communities to global markets and empowering individuals.

Secondly, the case study emphasizes the significance of cultural preservation and community engagement in resilience building. Igloolik has successfully utilized digital platforms to promote and preserve local Indigenous culture, fostering a sense of pride, identity, and intergenerational knowledge transfer. Other rural communities can take inspiration from this approach and leverage ICTs to celebrate their own cultural heritage, traditions, and unique assets. By embracing digital platforms for cultural preservation and community engagement, rural areas can cultivate a sense of belonging, foster resilience, and strengthen their collective voice amid challenges.

Lastly, the case study highlights the importance of addressing digital divides and investing in infrastructure upgrades. Rural areas often face limitations in connectivity and Internet access, hindering their full potential for resilience-building. Governments, organizations, and stakeholders need to prioritize infrastructure investments to ensure equitable access to ICTs in rural communities. By narrowing the digital divide, providing reliable and affordable Internet access, and promoting digital literacy, other rural areas can follow Igloolik's example and leverage ICTs as a catalyst for resilience, economic development, and social empowerment.

Like Igloolik, the Guna community in Panama faced urgent climate threats alongside digital exclusion. But here, the solution combined relocation, a deeply disruptive process, with technology and cultural agency in a way that offers lessons for adaptation efforts worldwide.

Case Study #5: Drone Mapping and Community Agency: A Case Study on Relocating the Guna Population in Panama

In Panama, the government and an Indigenous community have taken a unique approach to managed relocation in response to the impacts of climate change—co-creation accelerated by drones. Through inclusive forums and participatory decision-making processes, they sought to ensure that community members had agency in shaping their future amid environmental challenges. We compare this approach to similar efforts in the United States, examining the lessons that can be drawn and the potential for scaling such processes. By understanding the successes and limitations of climate-driven relocation initiatives, we can inform future strategies and advocate for more equitable and community-centered approaches to adaptation.

The case study examines the relocation efforts of the Guna, an indigenous tribe living on the sinking islands of the San Blas Archipelago off the coast of Panama. Facing the dual challenges of rising sea levels and overpopulation, the Guna community sought solutions to their sustainable relocation, a process sometimes referred to as managed retreat.

This case study highlights the role of drone mapping technology and the community's active agency in shaping the project (Crisis Response Journal, 2021).

Panama Flying Labs, hosted by the Universidad Tecnológica de Panamá, collaborated with the Guna community to assess available land for relocation and evaluate the current state of the islands. The project was initiated with funding from the Inter-American Development Bank. A requirement of the bank's involvement included emphasizing the importance of engaging the affected population in decision-making processes and gathering their insights. The project aimed to generate orthomosaics, images made from a construction of other images by using photogrammetry (photographic measuring) software to provide visual representations of the surveyed terrain. This was, in turn, used to understand better the land in both geographies, the space at risk, and the relocation site, to facilitate informed decision-making.

When Technology Meets Community Agency and Support

The Flying Labs team used drone mapping technology, specifically PIX4Dmapper, to conduct detailed surveys and land assessments. They were able to depict land characteristics, distances, and areas in great detail. The orthomosaics provided valuable information that could be readily interpreted by stakeholders, including Guna community members who may lack surveying expertise. This approach enhanced communication and enabled meaningful community participation, ensuring that their perspectives were considered during relocation planning.

Armed with these images, the Gardi Sugdub Committee, part of the Guna and a local community group from one of the islands of the same name, played a pivotal role in collaborating with the Flying Labs team. The committee helped make decisions and acted as liaisons (Newsome, 2023; Bojorquez & Breen, 2023). They guided the Flying Labs team through unmapped areas. They also kept the local population informed about the project's progress, which helped bolster trust among the groups. The remarkable support from the Guna community fostered a cooperative environment in which their local knowledge was shared and integrated into the mapping project. This active engagement ensured that the relocation project respected the Guna's cultural heritage and addressed their unique needs.

The project also incorporated technology transfer elements to ensure the work is sustainable and replicable. The lab team trained community members to use and interpret drone data. After the initial project was completed, the community could use the information to make decisions about their new home. They could also use those skills in employment or to do other projects. Meanwhile, the drone researchers learned about the San Blas' unique ecosystem and its challenges. Any of this applies to other communities and to other uses of technology and connectivity to address specific climate change issues. Alongside that is the exposure to Indigenous culture and learning how it connects to land use and climate mitigation, which will expand the field of research. The project itself should serve as an example of how community-based participatory action research should be community-centered participatory action research.

The next steps for the Guna are adjusting to life on the mainland of Panama. This includes continuing some livelihoods that are transferable, such as their intricate textile weaving, and developing others more suitable to their new home. This will take time and integration with the new technology and skills. In the meantime, the Guna are sharing their experiences and contributing to other communities' ecological planning.

Lessons for Applicability and Scalability in Other Regions

The process used in the relocation of the Guna population can serve as a valuable model for climate-change-induced relocations in other regions. This is particularly valuable for countries that are working with indigenous communities facing similar challenges. By employing drones to capture high-resolution imagery and spatial data, planners in urban and rural areas can gain valuable insights into the current state of the land. This will help identify and assess suitable relocation sites.

This information, similar to that in Panama, can be used to develop informed strategies for sustainable resettlement. One critical aspect of work with the Guna community that should be included in any "lessons learned" recommendation is the need to honor community knowledge and agency.

The involvement of local communities in decision-making processes is crucial. Recognizing the historical struggles and unique cultural identities of indigenous communities, it is essential to empower them to participate actively in planning and implementation. This includes respecting their traditional knowledge, perspectives, and priorities. By engaging indigenous communities as partners rather than passive recipients of assistance, the process becomes more inclusive, culturally sensitive, and effective in addressing their specific needs and aspirations.

Next, the utilization of drone mapping technology can be replicated in other indigenous communities facing similar challenges. Drones provide a cost-effective and efficient means of gathering high-resolution imagery and spatial data, enabling communities to assess and monitor environmental changes, plan for sustainable development, and make informed decisions. Sharing the knowledge and expertise gained from the Guna case study can help other indigenous communities understand the potential benefits of such technologies and adapt them to their unique contexts.

To ensure scalability, it is important to foster partnerships between local communities, academic institutions, non-governmental organizations, and government agencies. Collaborative efforts can facilitate the transfer of knowledge, resources, and technical skills, empowering indigenous communities to leverage innovative solutions tailored to their specific circumstances. The Guna relocation project exemplified the successful integration of drone mapping technology and community agency. By leveraging drone-based surveying and photogrammetry, the Flying Labs team facilitated informed decision-making and provided visual communication tools for stakeholders. The active participation of the Guna community throughout the project ensured that their perspectives, cultural values, and insights were incorporated into the relocation process. In promoting community agency, incorporating appropriate technologies, and fostering inclusive partnerships, the processes witnessed in the Guna population's relocation can serve as a blueprint for scaling similar initiatives in other indigenous communities facing ecological challenges.

Across all of these examples, the consistent thread is not the specific technology used, but the centrality of community priorities in design and implementation. This is the defining feature that turns tools into sustainable, equitable solutions.

Conclusion

We have explored inspiring solutions and case studies from communities at the forefront of change. From Broadmoor New Orleans' resilience journey to Anacostia's unique digital equity initiatives and the multiple benefits demonstrated in a Baltimore community's transition from divestment to sovereignty, we learned how connectivity can help urban communities accelerate their path to resilience. We also learned that the transformation is not merely digital; social connections also take on new contours.

In rural and Indigenous communities, we learned that connection to the Internet and technology not only offers solutions to the climate crisis but also an opportunity to expand outward. Igloolik's journey to build economic resilience and preserve cultural heritage helped the community adapt to climate change and also created a conduit for sharing its historical climate knowledge with researchers and other communities. In Panama, a remote community was not only connected but also actively used modern technology to co-manage a sustainable relocation project with researchers. The knowledge exchange continues to inform the work of both the community and the researchers. These case studies exemplify how to work with rural and Indigenous communities to overcome the digital divide and confront climate change.

They highlight the importance of community empowerment, collaboration, and the dismantling of systemic barriers in shaping a more equitable and sustainable future.

The lessons from these communities also point to a broader conclusion: the systems governing infrastructure, technology, and environmental risk must evolve to reflect the realities of a climate-impacted, digitally interconnected world. Recommendations are easily written, but difficult to design and implement, even with case studies, data, and other evidentiary support. As we seek solutions, it is clear that a "community-up" approach to designing a renewed social contract is the way to align governments, markets, and communities around the shared responsibility of building resilient and inclusive systems for the future.

References

Benton Institute for Broadband & Society. (n.d.). *When do we get our broadband?* https://www.benton.org/blog/when-do-we-get-our-broadband

Bojorquez, M., & Breen, K. (2023). Panamanian tribe to be relocated from coastal Island due to climate change: "There's no other option". *CBS News.* https://www.cbsnews.com/news/panamanian-tribe-to-be-relocated-from-coastal-island-due-to-climate-change-theres-no-other-option/

Broadmoor Improvement Association. (n.d.). Neighborhood association*: New Orleans, LA.* https://www.broadmoorimprovement.com

Climate Signals. (n.d.). *Hurricane Katrina (2005).* https://www.climatesignals.org/events/hurricane-katrina-2005

Clinton Foundation. (2015). Broadmoor lives on: How a community saved their New Orleans neighborhood. Medium. https://stories.clintonfoundation.org/bringing-sound-back-to-broadmoor-how-we-saved-our-new-orleans-neighborhood-8e3d54c5c81f

Crisis Response Journal. (2021). Rising sea levels: Relocating indigenous people in Panama. https://www.crisis-response.com/Articles/616229/Rising_sea_levels.aspx

Ford, J. D., Smit, B., Wandel, J., & MacDonald, J. (2006). Vulnerability to climate change in Igloolik, Nunavut: What we can learn from the past and present. *Polar Record, 42*(221), 127–138. https://doi.org/10.1017/S0032247406005122

Ford, J. D., et al. (2020). An update on Inuit perceptions of their changing environment. *Elementa: Science of the Anthropocene, 8*(1), 025. https://online.ucpress.edu/elementa/article/8/1/025

Haldane, V., et al. (2023). Social capital and resilience. *Humanities & Social Sciences Communications, 10*, 217. https://www.nature.com/articles/s41599-023-02217-5

Hall, G. M. (n.d.). *Placage, concubinage, and race in Louisiana*. Georgetown University Library Repository. https://repository.library.georgetown.edu/handle/10822/1051120

Harvard Kennedy School Case Program. (2008). *"Broadmoor Lives": A New Orleans neighborhood's battle to recover from Hurricane Katrina (A)*. https://case.hks.harvard.edu/broadmoor-lives-a-new-orleans-neighborhood-s-battle-to-recover-from-hurricane-katrina-a/

Isle de Jean Charles Resettlement. (n.d.). *About the resettlement*. https://isledejeancharles.la.gov/about-isle-de-jean-charles-resettlement

Merriam-Webster. (n.d.). *Divest*. https://www.merriam-webster.com/dictionary/divest

My New Orleans. (n.d.). *The making of Broadmoor*. https://www.myneworleans.com/the-making-of-broadmoor/

Newsome, M. (2023). How an indigenous community in Panama is escaping rising seas. *Science News*. https://www.sciencenews.org/article/indigenous-guna-panama-escape-rising-seas

Preservation Resource Center of New Orleans. (2016). Broadmoor neighborhood history. https://prcno.org/wp-content/uploads/2016/06/Broadmoor.pdf

Social Explorer. (n.d.). *COVID-19 and redlining analysis*. https://www.socialexplorer.com

The Canadian Encyclopedia. (2015). *Iglulingmiut*. https://www.thecanadianencyclopedia.ca/en/article/iglulik-inuit

The Undivide Project. (n.d.). https://www.theundivideproject.org

U.S. Green Building Council. (2014). *A tale of two neighborhoods*. https://www.usgbc.org/articles/tale-two-neighborhoods

U.S. News & World Report. (n.d.). *Richest counties in America*. https://www.usnews.com/news/healthiest-communities/slideshows/richest-counties-in-america

W8CED. (2023). *Our priorities*. https://w8ced.org/our%20priorities/

Ward 8 Community Economic Development. (n.d.). *A dream deserved: Realizing our collective emergence*.

WWNO. (2016). *The solution to Louisiana's coastal problem no one wants to talk about*. https://www.wwno.org/coastal-desk/2016-11-17/the-solution-to-louisianas-coastal-problem-that-no-one-wa

Young, J. C. (2019). Rural digital geographies and new landscapes of social resilience. *Journal of Rural Studies, 70*, 66–74. https://doi.org/10.1016/j.jrurstud.2019.07.001

Young, J. C. (2021). Environmental colonialism, digital indigeneity, and the politicization of resilience. *Environment and Planning E: Nature and Space, 4*(2), 230–251. https://doi.org/10.1177/2514848619898098

8

Building a New Social Contract for Climate and Digital Justice

Throughout this book, we have looked into the structural inequalities embedded in climate and digital systems, the economic frameworks that shape how resources and risks are distributed, and the technological and community responses emerging in response to these challenges, Taken together, these discussions point to a deeper institutional problem. Two forces, technological acceleration and ecological disruption, are reshaping the modern social contract. The governance systems that structured twentieth-century economic growth were not designed for a world characterized by planetary environmental change and ubiquitous digital connectivity.

From Analysis to Action

This final chapter moves from diagnosis to design. It argues that solving these intertwined crises requires rebuilding the social contract itself, a covenant that defines how power, responsibility, and resources circulate among governments, markets, and citizens. The twentieth-century contract, built on industrial growth, fossil energy, and centralized

M. C. Sanders, *Climate and Code*, https://doi.org/10.1007/978-3-032-28032-9_8

authority, cannot withstand the distributed realities of a warming, wired world (Raworth, 2017; Sassen, 2014).

As we enter the mid-2020s, the rollback of key U.S. programs, including the Inflation Reduction Act, Justice40, and multiple executive-order initiatives on broadband and environmental equity, has exposed the fragility of reforms rooted in a single administration's agenda (Eilperin & Grandoni, 2025). The rapid reversal of climate and broadband investments, after years of planning and early deployment, has sent a clear message to local governments, businesses, and communities: executive action alone is not enough. What endures must be structural. Legislation, institutional design, constitutional interpretation, and civic norms must embed climate and digital justice as baseline expectations of governance, rather than as optional experiments that can vanish with an election cycle (Sanders, 2024a; UNHRC, 2023). This chapter organizes that vision through three pillars: governance, corporate responsibility, and community agency. Each pillar revisits core ideas from earlier chapters, offers practical directions for implementation, and gestures outward to the international level. Together, they outline the architecture of a new social contract for a digital climate age.

Governments: From Episodic Policy to Enduring Governance

Governments, at every scale, remain central to any serious effort to align digital and climate systems with justice. History lessons from the United States government underscore how volatile public commitments can be when they depend on executive orders, discretionary grants, or branding initiatives rather than durable law. Programs launched under the 2021–2024 U.S. administration signaled a shift: Justice40 promised that a substantial portion of climate and infrastructure benefits would flow to disadvantaged communities; new broadband funds targeted unserved and underserved regions; and climate-related tax credits sought to accelerate decarbonization and green jobs. When a new administration unwound many of these efforts in 2025, the result was policy whiplash that discouraged long-term investment, fragmented state and local planning, and deepened public skepticism about federal reliability (Kettl, 2022; Eilperin & Grandoni, 2025).

Earlier chapters showed why that volatility is so costly. In Chapter 3, we saw that climate-vulnerable and digitally disconnected communities frequently overlap geographically. The same census tracts experience repeated flooding, heat stress, and pollution while also living with limited broadband access, unreliable cellular coverage, and fewer devices. The infrastructure that carries electricity, data, and water into neighborhoods rarely arrives in isolation. When policy treats climate resilience and digital access as separate domains, it misses chances to design solutions that address both at once. In Chapter 4, we argued that governments have treated equity as a discretionary add-on rather than as an investment that yields measurable returns in health, productivity, and fiscal stability (Raworth, 2017). Chapter 5 highlighted that technology policy is social policy because the rules for networks, procurement, and standards determine who can access emergency alerts, telehealth, online education, and democratic processes. Chapter 6 documented how communities self-organized when formal systems faltered, but also showed the limits of resilience when local action is not backed by sustained public support.

A more durable governmental role begins with law. Instead of relying on "once-an-administration" packages that expand and contract with partisan cycles, Congress and state legislatures can convert equity-oriented initiatives into statutory mandates. Climate and digital equity can become required considerations in infrastructure appropriations, akin to civil rights and environmental impact standards (EPA, 2022). Agencies that fund energy, transportation, housing, health, and telecom could be required to examine not only aggregate benefits but their distribution: who gains access, whose costs are reduced, and who remains excluded. In practical terms, that means integrating equity analyses into budget scoring, program design, and oversight.

To make this feasible across a fragmented federal system, the United States could establish a permanent coordinating authority, such as an Office of Resilience within the Office of Management and Budget. As the Council on Environmental Quality did for environmental review, this office would harmonize data standards and performance metrics across agencies, publish annual scorecards on digital access and climate resilience, and review large capital programs for their combined impacts (FAS, 2023; UNDRR, 2024). Instead of separate pipelines for broadband, flood control, and electrification, federal programs could be evaluated for their ability to deliver layered benefits: connectivity and decarbonization, hazard reduction, and administrative burden relief.

Governance must also become more participatory. Justice40 and similar efforts experimented with listening sessions and advisory councils, but these mechanisms have often remained consultative rather than decision-shaping. Converting temporary or ad hoc consultations into standing community advisory boards, with access to data and real influence over program priorities, shifts the relationship. For example, advisory bodies composed of community-based organizations, local governments, tribal nations, and social-service providers could review grant criteria, monitor project delivery, and publish their own evaluations of whether benefits actually reach intended populations (UNDP, 2022). Embedding local voices in rulemaking and budget oversight helps ensure that policies are grounded in lived experience rather than solely in model outputs or distant assumptions.

Better information is another essential ingredient. As we argued in earlier chapters, the United States lacks a comprehensive, interoperable system for disaster and recovery data. This is even more of an issue in less digitally developed countries. An Open Disaster Data statute could standardize how agencies collect, anonymize, and share hazard, loss, and aid data while protecting privacy. When communities, researchers, and journalists can trace how funds move, where damage is concentrated, and how quickly different neighborhoods recover, it becomes harder for inequities to remain invisible. It also becomes easier to improve programs over time (Lakhina et al., 2023).

Finally, governments must stabilize funding. Competitive, one-off grants encourage innovation, but they also demand substantial administrative capacity and favor applicants with sophisticated grant-writing teams. Small and rural communities, which often face the highest combined climate and digital risks, are least able to compete. Moving toward formula-based transfers, in which a portion of funds is distributed according to exposure, social vulnerability, and adoption outcomes, can establish a more predictable baseline. Multi-year appropriations and automatic stabilizers, funded through mechanisms such as spectrum auctions or carbon pricing, can further insulate core programs from political cycles (World Bank, 2023).

Taken together, these changes signal a shift from episodic policy to enduring governance. Digital access and environmental health become public goods that the state is obligated to protect, rather than discretionary targets of executive generosity. Law, institutional design, and open data build the scaffolding for accountability. Within that frame, other actors, especially corporations, can either reinforce or undermine the new social contract.

Corporations: From Extractive Enterprise to Shared Stewardship

Private firms control much of the infrastructure and innovation that shapes both digital and climate futures. Telecommunications providers, data center operators, cloud platforms, hardware manufacturers, and energy companies all wield significant influence over where networks are built, how energy and water are consumed, and which communities enjoy reliable access to essential services. The question for a new social contract is not whether corporations will be involved, but how they will be governed and what obligations they owe beyond compliance with minimal regulations.

Earlier, we described "technology as terrain." Those who build and control that terrain, whether through fiber routes, server farms, content moderation rules, and algorithmic design, shape who can move, speak, and participate (Sanders, 2024b). Chapter 4 challenged the familiar narrative that value creation must rest on extraction, whether of natural resources, labor, or data. We saw that business models based purely on externalizing costs, in both carbon and human terms, are increasingly incompatible with planetary limits and social stability (Raworth, 2017). Yet when federal climate programs were rolled back, many firms responded in predictable ways: some paused investments in decarbonization and community broadband, citing policy uncertainty; others used deregulation to expand or accelerate projects in ways that increased emissions, water stress, or local opposition (OECD, 2024; KPMG, 2022).

Moving from extraction to stewardship requires changes in both external rules and internal incentives. One part of that shift concerns disclosure. When securities law treats environmental and social performance as marginal "ESG" concerns, firms can under-report or selectively frame their impacts. Extending disclosure rules to include standardized indicators for emissions, water use, digital access rates, content governance, and algorithmic bias makes risks and responsibilities more legible to investors, regulators, and communities (SEC, 2023; OECD, 2024). That disclosure needs to be accompanied by reasonable assurance from independent auditors, so that it reflects reality rather than marketing.

Internal accounting must also change. Social return accounting can complement traditional financial metrics by tracking how a company's operations affect connectivity, resilience, and local economies. For example, a data center developer could evaluate not only construction costs and power purchase agreements, but also the share of local households that gain enhanced connectivity, the extent to which waste heat is used for district heating, and the reduction in emissions relative to a counterfactual site. Over time, tying executive compensation and project greenlight decisions to multi-year performance on these measures shifts incentives toward longer-term stewardship (KPMG, 2022).

In concrete terms, companies can invest in infrastructure that functions as a local common rather than as an isolated asset. Partnerships with municipalities and nonprofits to build community broadband networks, shared microgrids, public charging infrastructure, and community cooling centers can align corporate and public interests. When these projects are co-designed and include binding community-benefit agreements that cover local hiring, training programs, and governance, the benefits are more likely to endure (World Bank, 2023; Sanders, 2024b).

At the same time, firms need to rethink data practices. Many communities we encountered in the Ten States Project were acutely aware that data about them was being collected, either through social media platforms or through monitoring devices, without clear benefit-sharing or transparency. Responsible approaches would include privacy-preserving data collaboratives that enable local governments, researchers, and communities to use aggregated outage, mobility, and hazard-exposure data to improve service and planning without exposing individuals to harm (Lakhina et al., 2023).

Public policy can encourage, and sometimes require, this shift. Tax credits can reward verified gains in broadband adoption in marginalized areas, or demonstrable reductions in facility-level emissions and water use. Public procurement can prioritize vendors who meet clear standards for digital rights, environmental performance, and inclusive governance. Joint innovation hubs, co-funded by companies, universities, and philanthropy, can prototype new tools for equitable siting, water reuse, and just transitions, while open licensing ensures that publicly supported tools do not become proprietary choke points (World Bank, 2023; Sanders, 2024b).

In a new social contract, corporations are not purely philanthropic actors nor purely regulated entities. They are stewards of critical systems whose license to operate depends on aligning their core business decisions with equitable and sustainable outcomes. When those expectations are clear, markets can scale solutions rather than amplify harm. But market transformation alone is insufficient—the legitimacy and relevance of any social contract rest on the communities themselves.

Communities: From Impacted Populations to co-Designers of the Future

Communities sit at the junction of digital and environmental systems. They experience both the immediate shocks, storms, fires, floods, outages… and the slow violence of chronic underinvestment and exclusion. Historically, policy and corporate decisions have framed communities, especially marginalized ones, as impacted populations to be served, studied, or relocated. A social contract grounded in climate and digital justice reframes them as co-designers and co-governors.

Chapter 6 offered concrete examples. Neighborhood groups used participatory mapping and storytelling to document flooding, housing damage, cultural loss, and connectivity gaps in ways that statewide datasets and federal reports did not capture. In rural Iowa, interviews and data suggested that census tracts with the poorest broadband access were often those with high flood risk, proximity to CAFOs, or industrial pollution. Residents in these zones were less able to receive timely alerts, upload damage documentation, or participate effectively in planning processes. That dual deprivation of physical exposure and digital exclusion perpetuated cycles of underinvestment and distrust (Sanders et al., 2023).

These cases illustrate the importance of community-generated evidence and governance. When residents collect and interpret their own data, they can challenge official narratives that ignore local realities. Community dashboards that integrate hazard exposure, health outcomes, administrative burden, and connectivity can reveal patterns that justify claims for resources and change. If those systems are designed with clear protocols for consent, privacy, and access, they can also become durable tools for local decision-making (UNDP, 2022; Lakhina et al., 2023).

Case Example: AI, Disaster Prediction, and the Burden of Incomplete Data. In recent years, federal and state agencies have experimented with artificial intelligence to support disaster prediction, early warning systems, and damage assessments. These tools promise speed and precision. One example occurred after the 2023 Midwestern flood season, when an AI-assisted damage assessment tool was piloted to estimate household losses using satellite imagery, parcel data, and historical flood records. In higher-income suburbs, where property records were current, roof structures were uniform, and broadband allowed real-time reporting, the model performed well. In rural and low-income areas—especially those with limited broadband coverage—the system returned significantly lower loss estimates. Homes with unpermitted additions, manufactured housing, or older construction materials were misclassified. Debris fields, mold spread, and interior damage often went undetected because satellite imagery could not capture conditions inside the structures, and residents lacked the connectivity to upload photos or file digital reports promptly.

A state audit later found that AI-generated estimates in these communities understated damages by as much as 40 percent. Families received smaller awards, and local governments struggled to justify requests for infrastructure repairs. Because the model relied heavily on historical data, it also reproduced outdated hazard maps that failed to reflect recent shifts in rainfall and runoff patterns, precisely the types of changes climate scientists have warned about for decades.

This case illustrates why AI cannot operate effectively without inclusive, high-quality data and participatory governance. When residents lack the connectivity needed to report conditions, their absence from the dataset becomes misinterpreted as a lack of need. When infrastructure and housing vary widely, standardized imagery produces systematic blind spots. Without mechanisms for community validation, algorithmic predictions can become official truths, even when they contradict lived experience.

Several steps emerged from this failure. The state partnered with local libraries and extension offices to create upload hubs where residents could submit photos and reports. Community-based organizations were invited to help verify AI outputs before decisions were finalized. Satellite analysis was supplemented with ground-truthing done by local emergency managers familiar with construction patterns in their jurisdictions. Most importantly, the state revised its procurement guidelines to require equity impact assessments for all automated decision tools used in disaster management.

This example underscores one of the central themes of the book: AI systems inherit the inequalities of the environments in which they operate. When digital exclusion and environmental vulnerability overlap, those most exposed to harm are also most likely to be misrepresented by the tools intended to guide recovery. Ensuring fairness in AI-assisted disaster governance requires not only technical safeguards but also investments in connectivity, participatory processes, and ongoing evaluation. These are all elements that form the backbone of a just digital, and climate-ready society.

Supporting equity at the local level also requires metrics and accountability. Municipalities and counties can adopt regular "equity audits" that track how climate adaptation, disaster recovery, and connectivity investments are distributed across neighborhoods. Instead of reporting only total miles of fiber laid or total number of resilience projects funded, they can publish small-area data on adoption, latency, outage frequency, flood risk reduction, and heat exposure. Layering this information with demographic data makes it possible to see whether resources are reaching communities with the greatest need or reinforcing existing hierarchies (CDC, 2022; OECD, 2024).

Community resilience also depends on knowledge infrastructure. Libraries, schools, faith organizations, neighborhood associations, and cultural centers already serve as nodes of trust and information exchange. Some of these are already resilience hubs, places with backup power and connectivity. Other spaces are for climate, digital literacy, and emergency preparedness training. With support, these venues can expand to become places where residents can receive assistance with applications, appeals, and documentation that supports both day-to-day well-being and crisis response (Brookings Institution, 2023; IFRC, 2021). When these institutions are resourced to host community-generated data, convene cross-sector dialogues, and anchor local memory, they become key sites of co-creation.

Ownership is a further dimension. Community-based cooperatives and municipal broadband networks can, in some contexts, provide more reliable and accountable service than distant providers. Similar models can apply to energy, particularly for microgrids and distributed generation. Where cooperative or public ownership is not practical, communities can still negotiate community-benefit agreements that include clear performance benchmarks, equity stakes, and mechanisms for shared decision-making.

None of this can be achieved without resources. Locally governed resilience funds, capitalized by philanthropy, settlements, and dedicated public revenue, can provide flexible match funding and rapid-response capacity. Partnerships with universities can supply technical assistance, data stewardship, and evaluation. Simplified contracting and procurement rules can make it easier for small community-based organizations to serve as prime partners rather than perpetual subgrantees.

When governments create structural room for participation, and corporations open their models to accountability, communities can move from marginal voices to central partners. The question then becomes how to design systems that sustain this alignment over time.

Building Systems that Endure: Implementation and Integration

Sustainability is not only about emissions levels or green infrastructure. It is also about whether systems can persist and adapt when leadership changes, budgets tighten, or new risks emerge. Reforms that live or die by electoral outcomes may produce bursts of progress, but they cannot anchor a generational transition. A new social contract must be designed to endure.

Earlier chapters pointed to the temporal mismatch at the heart of modern governance. Hazards unfold over hours, seasons, and decades. Digital tools and business models evolve in months and years. Political attention spans are shorter still. Chapter 2 examined how technological speed can outpace democratic oversight. Chapter 5 suggested that decentralized technologies could, if governed well, embed a degree of permanence and resilience through redundancy and local control. The thread that ties these together is the need for institutions that learn, adjust, and persist.

One approach involves building intergenerational frameworks. Independent resilience funds or endowments, governed by boards that include representatives from government, community organizations, labor, and the private sector, can provide stable support for maintenance and upgrades to critical digital and climate infrastructure. Unlike project-based grants, endowments are designed to generate income in perpetuity. If structured transparently and with clear mandates, they can finance mundane yet essential tasks like replacing outdated equipment in a public cooling center or upgrading neighborhood networks, long after the initial construction ribbon-cutting has passed.

Standardization also plays a role. Shared metrics and data standards for digital inclusion, emissions intensity, and adaptation progress make it easier to compare performance across jurisdictions and over time (UNDRR, 2024). If every city or state defines "served" broadband, "resilient" infrastructure, or "equitable" allocation differently, accountability becomes difficult. By contrast, a common set of indicators, published as open data, allows residents, researchers, and civil society to benchmark progress and identify lagging areas.

Learning needs to be formalized. Rather than treating evaluation as a compliance exercise at the end of a grant period, agencies and partners can establish feedback loops from the beginning. Regular check-ins with communities, adaptive management of projects, and willingness to adjust course based on evidence help prevent rigid adherence to flawed designs (Smit & Wandel, 2006). Documenting what does not work and making that documentation public ensures that other jurisdictions do not repeat the same mistakes.

The design of administrative processes matters as well. As we saw in earlier chapters, administrative burden is itself a form of inequity. Complex applications, redundant reporting requirements, and incompatible systems push local governments and community organizations to their limits, especially in smaller or poorer jurisdictions (OECD, 2024). Harmonizing forms, aligning reporting cycles, and allowing secure data sharing across programs can reduce these burdens and make participation more feasible.

Funding mechanisms can be structured for stability. Dedicating a portion of spectrum auction revenues or a modest carbon fee to digital–climate infrastructure funds provides a recurring revenue stream that is less vulnerable to annual appropriations politics (World Bank, 2023). Introducing "sunrise" provisions, which require a formal review and replacement plan before key equity-oriented laws or programs can be repealed, also challenges the current pattern in which reforms expire while regressions are permanent.

Finally, institutional memory needs support. Libraries, archives, and universities can preserve plans, datasets, and community agreements so that each new cohort of officials and advocates does not have to rediscover basic facts or renegotiate long-settled principles. This cultural infrastructure helps sustain the social contract even when formal rules change.

These elements, intergenerational funds, common metrics, feedback loops, administrative simplification, stable revenue, and preserved memory, turn discrete initiatives into systems. They also prepare the ground for cooperation beyond national borders, because enduring domestic systems are necessary preconditions for credible international commitments.

Global Collaboration: Governance for a Just Internet and Digital Economy

Digital infrastructure and greenhouse gases both move across borders without regard for jurisdiction. The same platforms that deliver telemedicine and educational access can spread misinformation, amplify surveillance, or extract data from communities with little regulatory protection. Nations most exposed to climate extremes often have the least resilient infrastructure and the least leverage in global standard-setting forums. As a result, the burdens of the digital economy and the climate crisis converge unevenly, reproducing inequalities documented in earlier chapters (UNESCO, 2023; Sassen, 2014).

International governance has not kept pace with these realities. Climate agreements continue to frame emissions and adaptation as primarily environmental issues. Digital governance regimes, meanwhile, focus on trade, cybersecurity, and intellectual property rather than on equity or human rights. Artificial intelligence introduces a third layer of complexity. AI systems increasingly shape disaster response, credit scoring, public benefits eligibility, labor markets, and energy demand. Yet many countries lack the baseline digital infrastructure to ensure that AI systems function safely, transparently, and accountably (UNESCO, 2023; UNFCCC, 2024). Without a global framework that links climate, digital access, and automated decision-making, disparities in risk and opportunity will widen.

Earlier chapters traced how domestic patterns of marginalization scale globally. Chapter 3 showed that those least responsible for emissions often face the greatest climate exposure; Chapter 5 revealed parallel patterns in the digital arena, where countries rich in capital and connectivity set the terms of engagement, while others serve primarily as sites of resource extraction, data flows, or outsourced labor. These asymmetries limit the ability of many nations to participate in the digital economy on fair terms. They also constrain how quickly and safely AI systems can be deployed in climate adaptation, disaster management, and public administration.

Case Example: AI, Hurricane Response, and Data Inequity in the Caribbean. Caribbean governments and regional organizations have increasingly turned to artificial intelligence to support hurricane preparedness and response. After Hurricanes Irma and Maria, several pilot projects used AI models to predict storm intensity, map expected damage, and prioritize aid delivery. These models were trained using satellite imagery, shipping and aviation data, and historical storm tracks from U.S. and European agencies. While the tools improved forecasting accuracy along heavily monitored corridors, they struggled in areas where data infrastructure was limited or outdated.

A notable example came during the 2024 hurricane season, when an AI-assisted logistics platform was deployed to coordinate supply distribution across several Eastern Caribbean islands. The model relied on broadband-based reporting from local emergency operations centers, automated port-status updates, and crowdsourced data submitted through a mobile application. Larger islands with more robust fiber networks were able to provide real-time situational reports, and the platform routed supplies quickly to their designated distribution points.

Smaller islands, however, experienced rolling outages during the storm and had limited pre-storm connectivity to begin with. Their reports arrived sporadically or not at all. The AI system interpreted the absence of data as an indication that damage was minimal or that supply chains remained functional. As a result, the algorithm recommended prioritizing inventory for locations that were already better connected—and, in some cases, less severely impacted. By the time human responders realized the model's interpretation was inaccurate, several communities had gone days without adequate water, medical supplies, or fuel.

Local disaster officials later explained that the model had missed critical variables: informal ports used by fishers, unpaved roads washed out repeatedly during storm surges, and the presence of multi-generational households dependent on local clinics. These realities were not captured in the datasets used to train the AI. A post-season evaluation by the Caribbean Disaster Emergency Management Agency (CDEMA) concluded that "data deserts" within the region created systematic blind spots in AI-supported decision-making and recommended a regional data-governance framework that integrates community reporting channels, local GIS systems, and non-digital information streams.

This case underscores the central theme in this book: AI does not eliminate inequity; it reflects the data environment it inherits. When connectivity gaps map onto colonial-era infrastructure patterns and climate vulnerabilities, automated systems reinforce structural disadvantages rather than alleviating them. Building a just digital and climate-ready Caribbean future will require interoperable data standards, resilient local networks, and participatory validation mechanisms that ensure AI reflects the conditions on the ground—not just the conditions visible from satellites or major fiber routes.

A social contract fit for this century requires a multilateral framework that treats digital access, trustworthy AI, and climate resilience as interconnected global public goods. Doing so requires more than a treaty that focuses narrowly on connectivity or privacy. It requires a coordinated approach to the digital economy that includes infrastructure, rights, governance, and economic participation.

One model is a global agreement on a just digital economy. Such an agreement could build on elements of the Paris Climate Accord and pair nationally determined commitments on emissions with nationally determined commitments on digital justice. These commitments might include universal service benchmarks for connectivity; minimum standards for data protection; algorithmic transparency and auditability; commitments to reduce the carbon footprint of data centers; and requirements to support open, low-carbon technologies in both public and private sectors (UNFCCC, 2024; UNESCO, 2023).

Artificial intelligence must be included explicitly in this framework. AI governance remains fragmented, with competing standards emerging from the EU, the OECD, and industry coalitions. A global agreement could harmonize baseline protections: transparency requirements for high-impact models; safeguards against algorithmic discrimination; shared standards for data provenance and quality; and obligations for companies to assess and reduce the environmental impacts of model training, storage, and deployment (OECD, 2024). Climate-related AI tools should be subject to similar scrutiny, ensuring that predictive models used for disaster planning or insurance underwriting do not reproduce historical biases or penalize communities with poor infrastructure.

Equally important is the question of economic participation. The digital economy rewards countries that control capital, cloud infrastructure, and intellectual property. Countries without these assets risk being locked into digital dependency, purchasing expensive services from firms headquartered elsewhere. A just international digital economy would promote open standards, invest in shared infrastructure, and support local capacity to develop and govern digital tools. These measures would not only reduce dependence on foreign vendors but also expand domestic innovation and resilience (World Bank, 2023).

Climate finance and digital finance should be aligned rather than siloed. Investments in low-carbon energy, resilient water systems, and modern transportation are inseparable from investments in connectivity, data infrastructure, and trustworthy AI. Without reliable digital systems, nations cannot effectively monitor emissions, manage hazards, enable remote work, or participate in global commerce. Multilateral banks could condition certain loans on transparent governance of digital systems or offer concessional financing for green data centers and rural broadband. Sovereign wealth funds and philanthropic organizations could likewise support shared digital infrastructure that advances both climate and equity goals (UNDP, 2022).

Knowledge exchange remains essential. As we saw in earlier chapters, local innovations often arise in resource-constrained contexts. Open-source mapping tools, decentralized solar networks, and low-cost disaster information systems have emerged in climate-vulnerable regions out of necessity. When these innovations enter global circulation on equal terms, rather than as case studies that are admired but never adopted, they challenge the assumption that expertise flows only from wealthy nations to poorer ones (UNDRR, 2024; UNDP, 2022).

Institutional anchoring makes these commitments enforceable and durable. A Digital–Climate Justice Task Force housed within the UNFCCC could coordinate with the International Telecommunication Union, UNESCO, and regional bodies to align standards for connectivity, AI governance, digital rights, and environmental resilience (UNESCO, 2023; UNFCCC, 2024). Regional organizations such as the African Union, the Organization of American States, and ASEAN could adopt intergovernmental agreements that create pooled procurement for resilient digital infrastructure, establish shared data-governance principles, and coordinate cross-border hazard management.

International frameworks cannot substitute for domestic political will. Countries must build inclusive governance structures at home before they can credibly advocate for fairness abroad. At the same time, global frameworks help domestic reforms survive political turnover. They create a floor below which rights cannot easily fall and provide peer accountability when leaders attempt to dismantle critical protections.

A just Internet and digital economy, grounded in universal access, trustworthy AI, and climate-aligned infrastructure, represents an extension of the social contract into global space. It reinforces the idea that sovereignty includes responsibility: the responsibility to prevent harm, to share knowledge, and to ensure that technological and economic innovations serve human well-being across borders. It also clarifies the stakes of inaction. As digital systems and climate impacts grow more intertwined, inequitable governance of either system will reverberate worldwide.

This global dimension underscores the book's central claim: justice in a digital climate age cannot be achieved piecemeal. It requires alignment among laws, infrastructures, markets, and norms, and across local, national, and international scales. Global collaboration completes the circle by drawing the threads of community agency, governmental responsibility, and corporate stewardship into a shared global future.

Conclusion: Intentional Action and Shared Futures

The evidence gathered throughout this book leads to a simple but unsettling conclusion: the systems we inherited cannot carry us forward. The infrastructures that connect us, both digitally and physically, were built for extraction and expansion rather than for endurance and reciprocity. Rebuilding them requires both technical redesign and moral realignment.

A new social contract for the digital climate age rests on three commitments. The first is reciprocity. Access to clean air and water, to safe housing and infrastructure, and to trustworthy information and connectivity are reciprocal rights and responsibilities. Those who benefit from connected, low-risk environments have obligations toward those who do not. Connectivity must serve sustainability, and sustainability must sustain connectivity (UNHRC, 2023).

The second commitment is continuity. Governance cannot depend solely on political winds and executive preference. Laws, standards, and institutions must protect long-term justice from short-term agendas. Stable funding, multi-year planning, independent oversight, and protected data infrastructure can convert programs into systems that endure beyond any single leader or party (Kettl, 2022; UNDRR, 2024). Communities and corporations can plan, invest, and innovate more confidently when rules and commitments are not constantly at risk of reversal.

The third commitment is co-creation. The people most affected by environmental and technological disruption must help design the solutions. Community data initiatives, participatory budgeting, co-governed advisory boards, and locally owned infrastructure are not merely procedural niceties. They are essential mechanisms to ensure that knowledge flows upward as well as downward and that interventions respect local priorities and capabilities (UNDP, 2022; Sanders, 2024a).

The appendix that follows lays out model laws, regulatory frameworks, and institutional designs that can translate these commitments into practice. It is intentionally grounded and pragmatic. Yet the most important work lies beyond any policy blueprint. It resides in how societies, day by day, choose to balance power with participation, innovation with integrity, and speed with reflection.

If the industrial revolution asked how to produce more, and the digital revolution asked how to connect faster, then this era asks how to sustain together. To build the new social contract is to affirm that our fates are linked across wires and waters, clouds and climates. Justice is not a peripheral policy category. It is shared infrastructure. The act of rebuilding it will be the most enduring public work of our time.

References

Brookings Institution. (2023). Digital literacy and the future of inclusive growth. Brookings Press.

Centers for Disease Control and Prevention. (2022). *Environmental justice and health disparities*. CDC Public Health Reports.

Eilperin, J., & Grandoni, D. (2025, March 12). Rollback of climate initiatives illustrates fragile policy gains. *The Washington Post*.

Environmental Protection Agency. (2022). Clean air act: Summary of major provisions.

Federation of American Scientists. (2023). *Enhancing climate initiatives: Green jobs*. FAS Publications.

International Federation of Red Cross and Red Crescent Societies. (2021). Working with nature to reduce disaster risk. IFRC.

Kettl, D. F. (2022). *Bridgebuilders: Connecting policy across the political divide*. Brookings Institution Press.

KPMG. (2022). The future of ESG reporting: From disclosure to performance. KPMG Global Insights.

Lakhina, S. J., Gall, M., & Sanders, M. C. (2023). Increasing national resilience through an open disaster data initiative. *Federation of American Scientists*. https://fas.org/publication/increasing-national-resilience-through-an-open-disaster-data-initiative/

Organisation for Economic Co-operation and Development. (2024). ESG and digital governance: Trends and regulations 2024. OECD Policy Paper.

Raworth, K. (2017). *Doughnut economics: Seven ways to think like a 21st-century economist*. Chelsea Green Publishing.

Sanders, M. C. (2024a). Building a digital justice framework. *Federation of American Scientists*. https://fas.org/publication/building-a-digital-justice-framework-pef-monica-sanders/

Sanders, M. C. (2024b). Are environmental regulators around the world doing enough to ensure a just transition? *Illuminem*. https://illuminem.com

Sanders, M. C., Mustafa, N., & Gusman, H. (2023). The ten states project: Mapping climate risk and digital divide in the U.S. *South*. The Undivide Project.

Sassen, S. (2014). *Expulsions: Brutality and complexity in the global economy*. Harvard University Press.

Smit, B., & Wandel, J. (2006). Adaptation, adaptive capacity, and vulnerability. *Global Environmental Change, 16*(3), 282–292.

U.S. Securities and Exchange Commission. (2023). Proposed rule: Enhancement and standardization of climate-related disclosures. *Federal Register, 88*(45).

United Nations Development Programme. (2022). *Participatory data for resilient communities*. UNDP.

United Nations Disaster Risk Reduction. (2024). *Global risk assessment report 2024*. UNDRR.

United Nations Educational, Scientific, and Cultural Organization. (2023). Digital commons and universal connectivity. UNESCO.

United Nations Framework Convention on Climate Change. (2024). COP29 outcomes and digital-climate synergies. UNFCCC Secretariat.

United Nations Human Rights Council. (2023). *Resolution on the right to a clean, healthy, and sustainable environment*. UNHRC.

World Bank. (2023). *Financing inclusive connectivity and resilient infrastructure*. World Bank Group.

Appendix: Top Ten Recommendations From "Climate And Code"

1. **Expand Equitable Digital Infrastructure**
   Governments should treat broadband connectivity, satellite communications, sensor networks, and emergency communication platforms as essential public infrastructure. Public investment and targeted subsidies should prioritize rural regions, disaster-prone areas, and historically underserved communities where digital exclusion compounds climate risk.
2. **Build Digital Public Infrastructure for Climate and Disaster Resilience**
   National and regional governments should develop interoperable digital public infrastructure that supports environmental monitoring, emergency communication, data sharing, and public engagement. Climate adaptation plans, hazard mitigation strategies, and national resilience frameworks should explicitly incorporate digital connectivity and data systems as foundational resilience assets.

M. C. Sanders, *Climate and Code*, https://doi.org/10.1007/978-3-032-28032-9

3. **Develop Workforce and Community Technical Capacity for the Climate–Digital Economy**
   Public workforce programs and community education initiatives should integrate digital skills training with climate resilience and environmental management pathways. Governments and philanthropic institutions should support training in fields such as geographic information systems, remote sensing, environmental monitoring, data analytics, and artificial intelligence to enable communities to participate directly in resilience planning and emerging climate-related industries.
4. **Support Community-Led Data Systems and Participatory Monitoring**
   Governments and research institutions should fund participatory mapping, crowdsourced environmental monitoring, and community science initiatives. Integrating locally generated data into official risk models and resilience planning improves data accuracy, strengthens public trust, and ensures that local knowledge informs adaptation strategies.
5. **Establish Open and Accessible Climate and Infrastructure Data Systems**
   Public agencies should adopt open data standards that make climate, environmental, and infrastructure information widely accessible to researchers, communities, and policymakers. Open data platforms should include strong privacy protections and community governance mechanisms to ensure responsible use of shared information.
6. **Promote Ethical and Accountable Use of Artificial Intelligence in Climate and Disaster Systems**
   Governments should require transparency, bias testing, and independent oversight for artificial intelligence and machine learning systems used in disaster response, infrastructure planning, and climate risk modeling. Algorithmic audits and public reporting requirements will help ensure that digital tools produce fair and reliable outcomes.

7. **Protect Community Data Rights and Sovereignty**
   Legal frameworks should ensure that communities retain meaningful control over data generated through participatory monitoring, civic technology initiatives, and environmental data collection. Policies should support community data trusts and governance structures that manage data access, privacy protections, and benefit sharing.
8. **Encourage Public–Private Partnerships for Resilience Innovation**
   Governments should incentivize collaboration among public agencies, private firms, universities, and nonprofit organizations to develop technologies that support climate resilience. Priority areas include environmental monitoring systems, remote sensing, predictive analytics, and digital platforms that improve disaster preparedness and recovery.
9. **Support Technology-Enabled Nature-Based Solutions**
   Public investment in nature-based climate adaptation strategies should incorporate digital tools such as geographic information systems, environmental sensors, and satellite monitoring. Technology-enabled monitoring improves ecosystem management and strengthens the long-term effectiveness of restoration and climate resilience projects.
10. **Strengthen Community Institutions and Governance in Resilience Planning**
    Governments should support community institutions such as schools, libraries, cooperatives, and local organizations that provide digital access, environmental education, and resilience planning capacity. Policies should recognize community-led infrastructure models, including cooperative broadband systems and locally governed adaptation initiatives, and integrate these institutions into disaster planning and response systems.

Tools and Resources

Learn more about my **Social Innovator In Residence** journey at Grinnell College: https://theundivideproject.org/iowa-grinnell

– I had the honor of being the inaugural Social Innovator in Residence at Grinnell College, spending time on the campus and in the community, teaching as much as I learned. In this section of our website, you can see the community asset map that came from this collaboration, as well as a hazard mapping tool meant to support farmers and rural communities.

Story Map "Assets Versus Access: A Baltimore Internet Story": https://storymaps.arcgis.com/stories/1e6d5161d3c64d9c844a843993853859

– A walkthrough of how Baltimore went from a thriving industrial community to one intentionally excluded from technological advancement to a city using tech as part of its comeback.

The Ten States Project For Equitable Climate Risk: https://the-ten-states-project-for-equitable-climate-risk-undivide.hub.arcgis.com/

– Here you can find the dashboard representation of our research in the ten states in the Gulf South and Mid-Atlantic United States, the tools and maps we built, as well as more of our community partners' stories.

Kolibri 360 GmbH
Ferdinand-Jühlke-Straße 7
99095 Erfurt, Deutschland
produktsicherheit@kolibri360.de